High-Level Synthesis Hardware Trojan Attacks and Countermeasures

Other related titles:

You may also like

- PBCS084 | Sengupta | High-Level Synthesis-based Methodologies for Hardware Security, Trust and IP Protection | 2024
- PBCS080 | Sengupta | Physical Biometrics for Hardware Security of DSP and Machine Learning Coprocessors | 2023
- PBCS066| Chang and Cao | Frontiers in Hardware Security and Trust: Theory, design and practice | 2020
- PBCS067 | Sengupta | Secured Hardware Accelerators for DSP and Image Processing Applications | 2019
- PBCS076 | Sengupta | Frontiers in Securing IP Cores | 2020

We also publish a wide range of books on the following topics:
Computing and Networks
Control, Robotics, and Sensors
Electrical Regulations
Electromagnetics and Radar
Energy Engineering
Healthcare Technologies
History and Management of Technology
IET Codes and Guidance
Materials, Circuits, and Devices
Model Forms
Nanomaterials and Nanotechnologies
Optics, Photonics, and Lasers
Production, Design, and Manufacturing
Security
Telecommunications Transportation

All books are available in print via https://shop.theiet.org or as eBooks via our Digital Library https://digital-library.theiet.org.

IET MATERIALS, CIRCUITS, AND DEVICES 90

High-Level Synthesis Hardware Trojan Attacks and Countermeasures

Anirban Sengupta

The Institution of Engineering and Technology

About the IET

This book is published by the Institution of Engineering and Technology (The IET).

We inspire, inform, and influence the global engineering community to engineer a better world. As a diverse home across engineering and technology, we share knowledge that helps make better sense of the world, to accelerate innovation and solve the global challenges that matter.

The IET is a not-for-profit organization. The surplus we make from our books is used to support activities and products for the engineering community and promote the positive role of science, engineering, and technology in the world. This includes education resources and outreach, scholarships and awards, events and courses, publications, professional development and mentoring, and advocacy to governments.

To discover more about the IET, please visit https://www.theiet.org/

About IET books

The IET publishes books across many engineering and technology disciplines. Our authors and editors offer fresh perspectives from universities and industry. Within our subject areas, we have several book series steered by editorial boards made up of leading subject experts.

We peer review each book at the proposal stage to ensure the quality and relevance of our publications.

Get involved

If you are interested in becoming an author, editor, series advisor, or peer reviewer, please visit https://www.theiet.org/publishing/publishing-with-iet-books/ or contact author_support@theiet.org.

Discovering our electronic content

All of our books are available online via the IET's Digital Library. Our Digital Library is the home of technical documents, eBooks, conference publications, real-life case studies, and journal articles. To find out more, please visit https://digital-library.theiet.org.

In collaboration with the United Nations and the International Publishers Association, the IET is a Signatory member of the SDG Publishers Compact. The Compact aims to accelerate progress to achieve the Sustainable Development Goals (SDGs) by 2030. Signatories aspire to develop sustainable practices and act as champions of the SDGs during the Decade of Action (2020–30), publishing books and journals that will help inform, develop, and inspire action in that direction.

In line with our sustainable goals, our UK printing partner has FSC accreditation, which is reducing our environmental impact on the planet. We use a print-on-demand model to further reduce our carbon footprint.

Published by The Institution of Engineering and Technology, London, United Kingdom

The Institution of Engineering and Technology (the "**Publisher**") is registered as a Charity in England & Wales (no. 211014) and Scotland (no. SC038698).

First published: 2025

The Institution of Engineering and Technology
Futures Place,
Kings Way,
Stevenage,
Herts, SG1 2UA,
United Kingdom

www.theiet.org

British Library Cataloguing in Publication Data
A catalogue record for this product is available from the British Library

ISBN 978-1-80705-067-2 (hardback)
ISBN 978-1-80705-068-9 (PDF)
ISBN 978-1-80705-069-6 (ePub3)

Typeset in India by MPS Limited

Cover Image: MF3d/E+ via Getty Images

Contents

About the author

Anirban Sengupta (SMIEEE, FIET, FBCS, FIETE, PhD) is a full professor in the Department of Computer Science and Engineering at the Indian Institute of Technology (IIT) Indore. He has more than 340 publications and patents, including seven books, to his credit.

He is the recipient of several awards and honors, including Distinguished Contributor of IEEE Computer Society, Distinguished Lecturer of IEEE Consumer Technology Society, Distinguished Visitor of IEEE Computer Society, Eminent Speaker of ACM India, fellow of the IET, fellow of the British Computer Society, fellow of the IETE, IEEE Chester Sall Memorial Consumer Electronics Award, IEEE CESoc Outstanding Editor Award, IEEE CESoc Best Research Award from CEM, Best Research Paper Award in IEEE ICCE 2019, IEEE Computer Society TCVLSI Outstanding Editor Award, and IEEE TCVLSI Best Paper Award in IEEE iNIS 2017. He held/holds around 20 editorial positions on IEEE and IET journal boards. He is consistently ranked in Stanford University's Top 2% Scientists globally.

Further details are available at: http://www.anirban-sengupta.com/index.php

Part I

Foundations

Chapter 1

Introduction to hardware Trojans and high-level synthesis

Anirban Sengupta[1]

1.1 What is a hardware Trojan?

A hardware Trojan (HT) is a deliberate and malicious modification introduced into an integrated circuit (IC) or system-on-chip (SoC) during its design or manufacturing process, often with the intent to compromise their functionality, reliability, security, or confidentiality by an adversary (Chakraborty *et al.*, 2009). These modifications may take the form of additional logic, altered connections, or subtle changes in functionality, often designed to remain hidden until triggered by a specific event or input pattern. Unlike traditional software-based attacks (which can be patched or updated post-deployment), hardware Trojans are embedded deep within the physical design, making them extremely difficult to detect once fabricated. Their stealthy nature allows them to stay dormant for long periods, thereby evading conventional validation and tests (Liu and Zhang, 2016).

The globalization of the semiconductor supply chain has amplified this threat, as modern hardware development relies heavily on third-party intellectual property (IP) cores, electronic design automation (EDA) tools, offshore fabrication facilities, and contract design services (Anshul and Sengupta, 2023a; Rostami *et al.*, 2014; Sengupta and Bhadauria, 2015, 2016). While this distributed design ecosystem accelerates innovation and reduces development cost, it also introduces multiple untrusted points where adversaries can secretly embed hardware Trojans. Each external dependency creates potential insertion points where untrusted entities may implant Trojans without immediate detection. Since these modifications may remain dormant until triggered under rare or specific conditions, they evade conventional verification and testing methods. Consequently, hardware Trojans represent a serious security challenge, raising concerns about the trustworthiness of devices that form the backbone of critical infrastructures, consumer electronics, and defense systems, where hardware integrity directly influences national security and user privacy (Rostami *et al.*, 2014; Dong *et al.*, 2020; Hu *et al.*, 2021).

[1]Department of Computer Science and Engineering, Indian Institute of Technology Indore, India

What makes hardware Trojans particularly concerning is their wide range of possible effects. Some Trojans may degrade performance or subtly alter output, while others may leak confidential information or disable functionality under specific conditions (Sengupta *et al.*, 2024). Because the triggers are often rare or cleverly disguised, traditional design verification and testing strategies struggle to detect them. This has elevated hardware Trojan detection and prevention into a central challenge in the hardware security research domain. In practice, ensuring trust in modern hardware requires rethinking design methodologies, integrating security mechanisms at higher levels of abstraction, and carefully monitoring the supply chain to mitigate Trojan risks before chips are deployed in the field (Rajendran *et al.*, 2013). Figure 1.1 highlights the complete hardware design flow and potential hardware vulnerabilities in each stage. The green components (shown in Figure 1.1) represent the trusted sources, where no hardware threat is possible, while red and yellow components represent the untrusted sources, where hardware threats are possible. Moreover, Figure 1.2 illustrates the different Trojan attack scenarios in the integrated circuit (IC) design and development cycle. In the first scenario (case 1), a third-party IP (3PIP) vendor is viewed as untrustworthy, while the SoC integrator and foundry

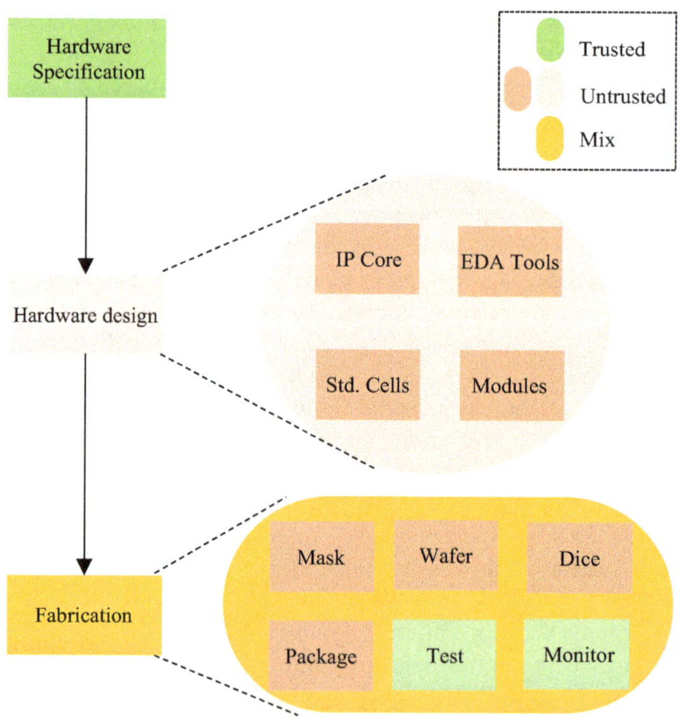

Figure 1.1 Hardware design flow and potential hardware vulnerabilities in each stage

	Case 1	Case 2
3PIP vendor	Attacker	Protector
SoC integrator	Protector	Protector
Foundry	Protector	Attacker

Figure 1.2 Trojan attacks in the hardware (IC) design cycle

are assumed secure. To counter the potential threat of backdoor Trojan insertion in the IP design [using the computer-aided design (CAD) tool library], the SoC integrator can enforce security measures during SoC design, such as assigning hardware from different vendors for critical operations or applying code coverage analysis with specialized test vectors to detect unusual signals that may indicate hidden Trojans. In the second scenario (case 2), the foundry is the weak link, where attackers may insert a Trojan during fabrication. Techniques like path delay analysis, thermal profiling, or hybrid testing approaches are used to expose such threats (Rostami *et al.*, 2014; Rajendran *et al.*, 2013).

1.2 Classification of hardware Trojans and their characteristics

Figure 1.3 illustrates the classification tree of hardware Trojans. As shown in Figure 1.3, hardware Trojans are generally classified into two main categories: *invasive* and *non-invasive*. *Invasive threats* involve direct modifications to the circuit and are most commonly introduced as digital hardware Trojans. *Non-invasive threat*s, on the other hand, do not alter the circuit physically but exploit external mechanisms such as negative bias temperature instability (NBTI) stress, which accelerates transistor aging, ultimately degrading performance and reducing the device's lifetime (Sengupta, 2017a, 2017b). Within invasive digital hardware Trojans, classification is further refined into *trigger-based* and *payload-based* categories. Trojans, when classified in terms of trigger, can be described in terms of triggering mechanism, *viz*., a rare combinational condition or a specific sequential event, which makes them extremely stealthy and difficult to detect during standard verification; while Trojans, when classified in terms of payload, can be described in terms of the malicious actions that they execute on triggering, which may include altering the functional behavior of a circuit, leaking confidential data, degrading performance without changing functionality, or causing denial of service, etc. (Sengupta, 2017a, 2017b).

The characteristics of hardware Trojans add another layer of complexity to their detection. Their triggers are often embedded in seldom-used signals or disguised as normal control states, effectively deceiving verification processes. Many Trojans are designed to activate only under rare conditions, which makes exhaustive testing impractical. Furthermore, some may only corrupt a small

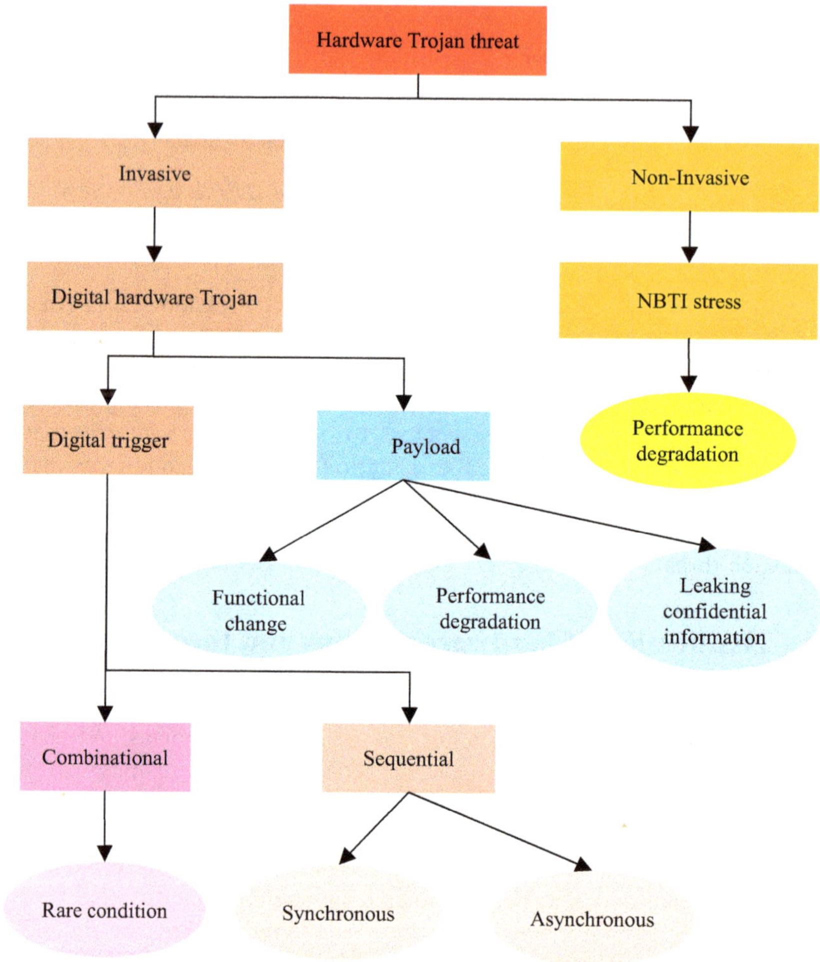

Figure 1.3 Classification tree of hardware Trojan threat

subset of outputs, while others continue to produce correct results, thus blending into the system's normal behavior. Standard techniques such as register transfer level (RTL) verification with targeted test vectors provide limited assurance and cannot guarantee complete Trojan detection, especially in the case of large third-party intellectual property (3PIP) cores that are often assumed pre-verified and bypass rigorous scrutiny (Sengupta, 2017a, 2017b). These classifications and characteristics highlight why hardware Trojans are considered one of the most insidious and challenging threats in modern consumer electronics design, demanding advanced detection strategies and security-aware design methodologies.

1.3 Taxonomy of HLS-based hardware Trojans

The taxonomy of hardware Trojans in high-level synthesis (HLS)-based IP design can be understood by examining the various covert insertion points within the design cycle. The complete hardware Trojan taxonomy tree, including the Trojan insertion design stage, Trojan types, payload, and detection techniques, is illustrated in Figure 1.4 (Sengupta, 2017a). As shown in Figure 1.4, Trojans may be injected during multiple stages, each introducing a distinct threat type. In the HLS tool library, an attacker may insert a *Functional Hardware Trojan (F-HT)* that directly alters output functionality. At the interconnect design stage, an adversary can create a *Denial-of-Service Hardware Trojan (DoS-HT)*, which

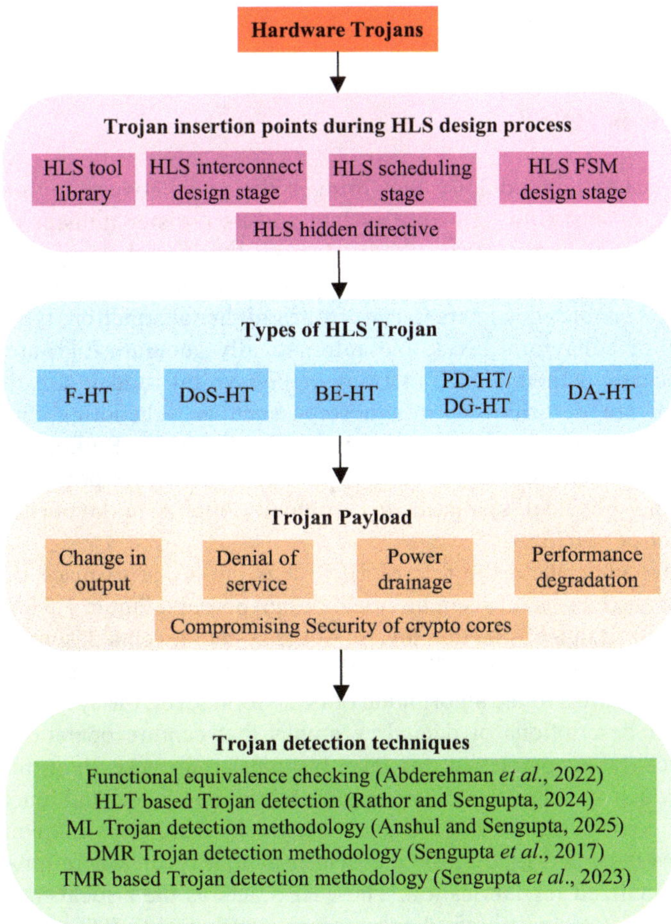

Figure 1.4 Taxonomy of HLS hardware Trojans

blocks or disables execution. At the interconnect design stage and finite-state machine (FSM) design stage, covertly inserted malicious logic can lead to *Performance Degradation (PD-HT)* or *Degradation Hardware Trojan (DG-HT)*, both of which subtly reduce the system's speed or efficiency without altering its functional correctness. During the scheduling phase, an adversary may deploy *Battery Exhaustion Hardware Trojan (BE-HT)* that drains power by forcing unnecessary computations. Finally, hidden Trojan directives in the HLS design process can give rise to *Downgrade Attack Trojan (DA-HT)*, which compromises cryptographic security by reducing the number of operational rounds in crypto-cores such as AES or SHA (Sengupta *et al.*, 2024; Rajendran *et al.*, 2013; Pilato *et al.*, 2024; Sengupta, 2017a, 2017b). Each of these Trojan types is associated with a different malicious payload, ranging from functional corruption and denial of service to excess energy dissipation and weakened security, thereby demonstrating the diverse and stealthy nature of Trojan taxonomy in IP designs.

1.4 What is HLS?

HLS is a design methodology that bridges the gap between high-level algorithmic descriptions of a hardware system and its register transfer level (RTL) implementation (Coussy *et al.*, 2009). Unlike traditional design flows where engineers manually code RTL descriptions in languages such as VHDL or Verilog, HLS enables designers to start from a higher abstraction, typically at the algorithmic or behavioral level, and automatically generate RTL architectures. In this process, mathematical functions or behavioral models are transformed into a hardware description through steps such as scheduling, binding, and resource allocation (Anshul and Sengupta, 2023a, 2023b; Rathor *et al.*, 2024). The result is an optimized RTL design that can be further processed by logic synthesis and physical synthesis to eventually realize a layout-level design (Sengupta *et al.*, 2023).

To better understand the role of HLS, it is important to place it within the hierarchy of abstraction levels in the VLSI design process. Figure 1.5 highlights the different VLSI design abstraction levels. At the system level, a design is defined in terms of inputs, outputs, and functional requirements such as performance and power. Moving down to the algorithmic or behavioral level, the system is expressed as high-level descriptions or data flow graphs that capture operations and their dependencies. HLS operates at this level by translating these algorithmic models into RTL, which specifies the transfer of data between registers and functional units under clock control. Below RTL lies the logic level, where designs are mapped into gate-level netlists, and finally, the physical or layout level, where transistor-level details are finalized for fabrication. Thus, HLS acts as the critical transformation stage that connects algorithmic description to synthesizable RTL, enabling faster exploration of design alternatives while reducing manual intervention (Anshul and Sengupta, 2023b; Sengupta *et al.*, 2023).

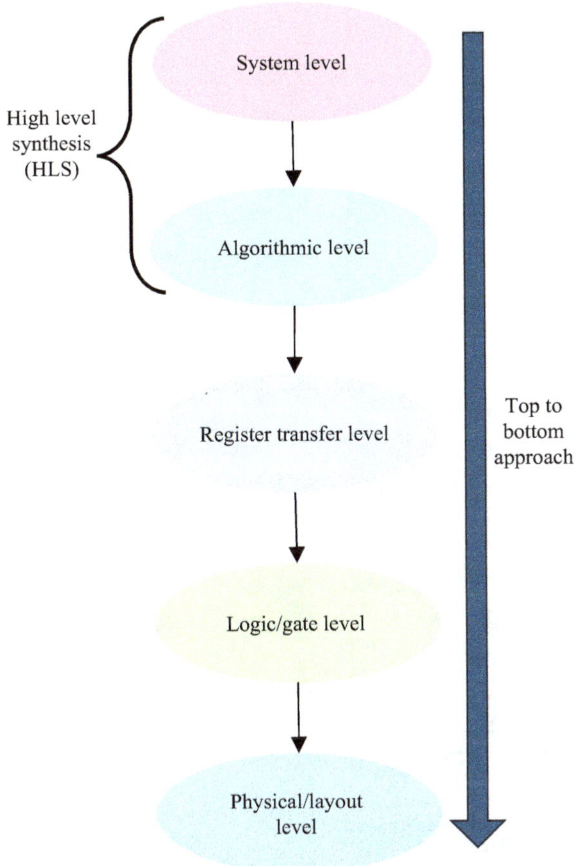

Figure 1.5 VLSI design abstraction levels

Next, Figure 1.6 illustrates the RTL optimization design flow. As shown in Figure 1.6, the system takes an RTL model of an electronic design and maps it into an efficient high-level hierarchical representation of the hardware implementation of the design. The hardware implementation is partitioned into functional partitions through automatic partitioning. This creates a fully characterized performance envelope for a range of feasible implementations for each of the partitions. The next step, called chip-level optimization, yields compact globally routed floorplans by choosing and refining the physical implementations of the partitions produced before. This process repeatedly invokes re-partitioning passes in order to refine the partitions and also calculates the suitable implementations. This is how a multiple-pass process converges on an optimal selection of physical implementations for all partitions for the whole chip. This is done while taking into account the minimum timing requirements and other design goals and constraints. The system finally produces particular control and data files that define the implementation details of the design throughout the

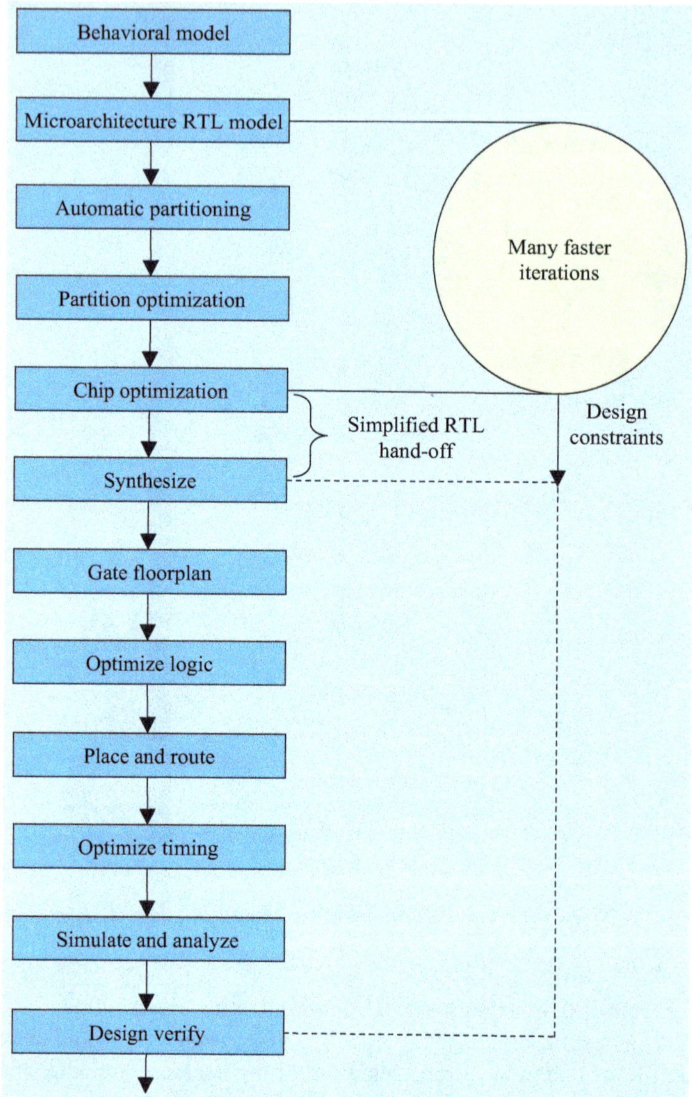

Figure 1.6 RTL optimization design flow diagram

whole backend flow process. This ensures that the final fabricated design meets all the design goals without expensive and time-consuming design iterations.

The importance of HLS has grown significantly in modern semiconductor design. As applications such as machine learning, digital signal processing, and multimedia processing demand high performance under tight power and area constraints, HLS provides a systematic way to explore trade-offs early in the design cycle (Sengupta *et al.*, 2023). By allowing rapid prototyping and automated generation of hardware architectures, HLS reduces design effort, shortens time-to-market, and enables the integration of advanced optimization or security features at higher abstraction levels. This makes HLS not only a productivity-enhancing tool but also a powerful framework for embedding trust and robustness in today's globally distributed hardware design ecosystem (Anshul and Sengupta, 2023a, 2023b; Rathor *et al.*, 2024; Sengupta *et al.*, 2016, 2023; Pilato *et al.*, 2018; Sengupta and Rathor, 2020, 2021).

1.5 Advantages of HLS

The adoption of HLS brings several key advantages to hardware design, making it a cornerstone of modern system-on-chip (SoC) development (Anshul and Sengupta, 2023a; Coussy *et al.*, 2009; Gajski and Ramachandran, 1994).

- **Design productivity**: One of the most significant benefits is design productivity. Since HLS allows designers to specify functionality at the algorithmic level, complex systems can be described in fewer lines of code compared to RTL, reducing development time and minimizing human error. This higher-level representation also makes it easier to modify or extend designs, facilitating iterative refinement and experimentation without the overhead of rewriting low-level hardware descriptions.
- **Design space exploration (DSE)**: Another advantage lies in DSE. HLS tools allow designers to quickly evaluate multiple hardware implementations of the same algorithm by varying constraints such as latency, area, and power. For instance, an algorithm can be scheduled to use fewer resources with higher latency or more resources with lower latency, depending on application requirements. This automated exploration capability ensures that designers can make well-informed trade-offs and arrive at optimal design points efficiently. Moreover, these explorations can be integrated with optimization frameworks such as evolutionary algorithms or swarm intelligence, further enhancing the ability to generate cost-effective and secure architectures. Figure 1.7 highlights the design space exploration process with the help of the NSGA-II core algorithm (Ferrandi *et al.*, 2008; Deb *et al.*, 2002). The authors in (Ferrandi *et al.*, 2024) have presented a methodology for design space exploration based on a multiobjective genetic algorithm. This work is based on the co-operation between the multiobjective non-dominated sorting genetic algorithm (NSGA-II (Deb *et al.*, 2002)) and a high-level synthesis design flow. The authors have proposed the following procedure for design space exploration: (a) the application of a multiobjective algorithm to automatically obtain a set of design solutions representing the trade-offs for the design objectives. (b) The introduction of a complete high-level synthesis to evaluate each design solution, considering all the components in the structural description. (c) Solution encoding suitable for area optimization, in particular when multi-modal functional units are involved in the design. The work in (Ferrandi *et al.*, 2008) is a complete methodology that accepts a C-based behavioral description of the problem as input, a library of resource descriptions, and a set of resource

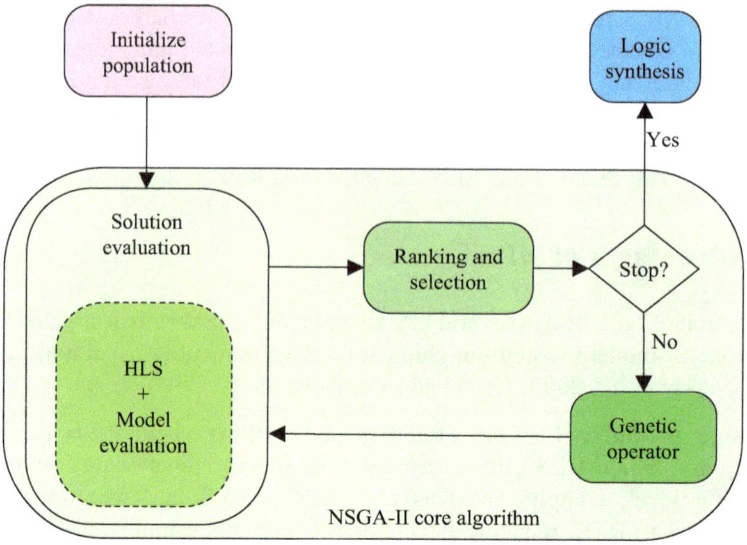

Figure 1.7 Design space exploration flow diagram

constraints specified in XML format. The framework then exploits NSGA-II, a multiobjective GA, to perform the design space exploration and concurrently optimizes the different design objectives.

- **Embedding security and reliability**: HLS also plays a crucial role in embedding security and reliability as design objectives into hardware IP designs. Since the design is captured at a higher abstraction, security constraints, such as watermarking, fingerprinting, or Trojan-resistant mechanisms, can be introduced systematically before the RTL stage. This ensures that the resulting RTL and subsequent lower-level designs inherit these security features, reducing the risk of piracy, tampering, or malicious modifications in the supply chain. Additionally, the automated synthesis process reduces manual RTL coding errors, improving design correctness and reliability.
- **Portability and maintainability**: HLS offers significant benefits in terms of portability and maintainability. Algorithmic descriptions are more reusable and portable across different platforms and fabrication technologies compared to low-level designs. This flexibility is particularly valuable in today's globalized semiconductor industry, where designs may need to be targeted for both FPGA prototyping and ASIC implementation. By abstracting the hardware design process, HLS accelerates innovation, supports heterogeneous system integration, and ensures that performance, cost, and security goals can be met simultaneously.

1.6 Security vulnerabilities in HLS flow

As discussed in the previous section, HLS has become an indispensable part of modern hardware design, enabling efficient translation of algorithmic specifications into RTL implementations. While HLS significantly improves productivity and accelerates custom IP design, it also introduces critical security vulnerabilities (Sengupta *et al.*, 2017, 2024, 2025; Pilato *et al.*, 2019; Anshul and Sengupta, 2025; Sengupta and Bhui, 2025; Abderehman *et al.*, 2022; Rathor and Sengupta, 2024). Since HLS involves multiple stages, such as scheduling, resource allocation, binding, interconnect design, and datapath/controller generation, each stage can act as a potential entry point for adversaries to insert backdoor hardware Trojans. For instance, compromised third-party HLS tools may introduce stealthy modifications in scheduling or allocation or even exploit unused ports in multiplexer-based design interconnects, leading to hidden backdoors in the final hardware IP (Sengupta *et al.*, 2024; Pilato *et al.*, 2019). Such Trojans may remain dormant under normal operating conditions and only activate under rare triggers, causing effects ranging from subtle performance degradation to complete denial of service.

HLS framework, many times, unintentionally generates unused ports (input and output ports in RTL components) in the final RTL datapath, which attackers can exploit for Trojan insertion. Recent studies have shown that malevolent HLS frameworks can exploit these vulnerabilities to inject Trojans within generated IP designs at minimal area and power overhead (Sengupta *et al.*, 2017, 2024, 2025; Pilato *et al.*, 2019; Anshul and Sengupta, 2025; Sengupta and Bhui, 2025; Abderehman *et al.*, 2022; Rathor and Sengupta, 2024). These attacks are particularly concerning because they bypass many traditional validation and detection strategies, making them extremely difficult to identify before deployment.

1.7 Academic and industry HLS tools

A variety of HLS tools, both academic and commercial, are currently used in the research and industrial community. On the academic side, *Bambu* provides a research-friendly, modular environment where algorithmic descriptions in C/C++ can be transformed, optimized, and converted to RTL, including features like hardware/software co-simulation, interface generation, loop unrolling, pipelining, and other architectural transformations (Ferrandi *et al.*, 2021). Similarly, *GAUT* generates controller and datapath architectures from C/C++ input and produces VHDL/SystemC RTL and memory/communication interfaces (Coussy *et al.*, 2008). Next, *LegUp* offers FPGA-targeted HLS with strong hardware/software co-design support (Canis *et al.*, 2011), and *SPARK HLS* is optimized for control-intensive designs and explores advanced scheduling and optimization strategies (Gupta *et al.*, 2003). Additionally, some academic HLS watermarking tools (such as the Crypto-steganography tool, the KHC-stego tool, and the Faciometric hardware security tool), which are primarily used to generate the watermarked RTL designs, include (CAD for Assurance, 2025a; CAD for Assurance, 2025b; CAD for Assurance, 2025c).

In industry, *CyberWorkBench* (NEC) is one prominent example: an "All-in-C" computer-aided design (CAD) suite with behavioral synthesis, formal verification, cycle-accurate and untimed simulation, interface generation, and optimization knobs for clocking, pipelining, memory, etc. (Wakabayashi and Schafer, 2008). Another widely used commercial suite is *Xilinx's Vivado HLS/Vitis HLS*, which allows designers to write functions in C or C++ (or use SystemC) augmented with pragmas/directives, simulate at a high level, synthesize to RTL targeting programmable devices, and fine-tune for area, latency, and interface type. It includes libraries for fixed-/floating-point arithmetic, support for various interface standards, and tools for analysis, debugging, and hardware/software integration (Xilinx, 2020). Additionally, *Cadence Stratus HLS*, *Mentor Catapult HLS*, and *Intel HLS Compiler* (from Intel) enable designers to write high-level algorithmic code with pragmas/directives and generate RTL optimized for latency, area, and throughput (Cadence Design Systems, 2015; Mentor Graphics/Siemens EDA, 2004; Intel Corporation, 2018).

These academic and industry HLS tools differ in their goals: academic tools often emphasize flexibility, new optimization algorithms, benchmarking, security as a design objective, quality of results (QoR), and open-source extensibility, while commercial tools focus more on robustness, ease of use, support, QoR, and certification (e.g., for safety-critical industries). Along with this, industry tools tend to have more mature front-end support, IP libraries, verification capabilities, and better user interfaces, whereas academic tools offer environments for prototyping novel HLS research, exploring watermarking and Trojan resistance, security enhancements, etc.

1.8 Additional utility of HLS—exploiting HLS for structural transformation and obfuscation

Structural obfuscation, from an IP designer's perspective, involves intentional modification to a hardware design's structure without affecting its true/desired functionality. This is to make it harder for an adversary (that may be present in untrustworthy foundries) to understand and analyze the underlying functionality of the design. This technique enhances hardware security by (a) countering reverse engineering, (b) mitigating hardware attacks, and (c) securing against IP piracy. Some of the prominent HLS-based obfuscation mechanisms are discussed in the subsequent section. The taxonomy of different HLS-based hardware obfuscation techniques in terms of key features/properties is shown in Table 1.1. Let us now look into some of the HLS obfuscation approaches.

The HLS obfuscation approach in Sengupta *et al.* (2019) introduces a methodology to secure DSP cores against common hardware threats, such as reverse engineering and counterfeiting/cloning. The key inputs for this methodology include (a) the DSP application in the form of C-code/mathematical functions or their equivalent DFG, (b) resource constraints, (c) module library, (d) structural obfuscation keys, and (e) the IP vendor-specified multivariable signature for the watermarking of the generated obfuscated design. Firstly, the methodology (as shown in Figure 1.8) involves applying structural obfuscation algorithms to the

Table 1.1 Comparison of different HLS-based obfuscation techniques in terms of various features/properties

HLS-based obfuscation techniques/Features of the approach	**Algorithm-based structural transformation**	**Usage of keys during HLS**	**Obfuscation of control branches**	**Design partitioning-based structural transformation**	**Inserting checkpoints and applicable for fault-secured IP designs**	**Production of sub-CDFGs due to structural transformation**	**Integrated design space exploration module**
C-way partitioning-based HLS obfuscation, 2024 (Sengupta and Chaurasia, 2024)	X	√	X	√	X	√	√
Loop folding-based structural HLS obfuscation, 2015 (Lao and Parhi, 2015)	√	√	X	X	X	X	X
Multi-cut based architectural HLS obfuscation, 2024 (Chaurasia and Sengupta, 2023)	√	X	X	X	√	√	X
HLS-based Key obfuscated RTL IP, 2021 (Islam *et al.*, 2020)	√	√	X	X	X	X	X
Multi-key based structural HLS obfuscation, 2020 (Sengupta and Rathor, 2020)	√	√	X	X	X	X	X

(Continues)

Table 1.1 (Continued)

HLS-based obfuscation techniques/Features of the approach	**Algorithm-based structural transformation**	**Usage of keys during HLS**	**Obfuscation of control branches**	**Design partitioning-based structural transformation**	**Inserting checkpoints and applicable for fault-secured IP designs**	**Production of sub-CDFGs due to structural transformation**	**Integrated design space exploration module**
Hologram-based structural HLS obfuscation, 2019 (Sengupta and Rathor, 2019)	X	√	X	X	X	X	X
Compiler/HLS transformations integrated with key management, 2018 (Pilato *et al.*, 2018)	√	√	√	X	X	X	X
Multilayered HLS-based algorithmic transformation, 2017 (Sengupta *et al.*, 2017)	√	X	X	X	X	X	√
HLS-based functional obfuscation with optimized Key, 2019 (Sengupta *et al.*, 2019)	X	√	X	X	X	X	√

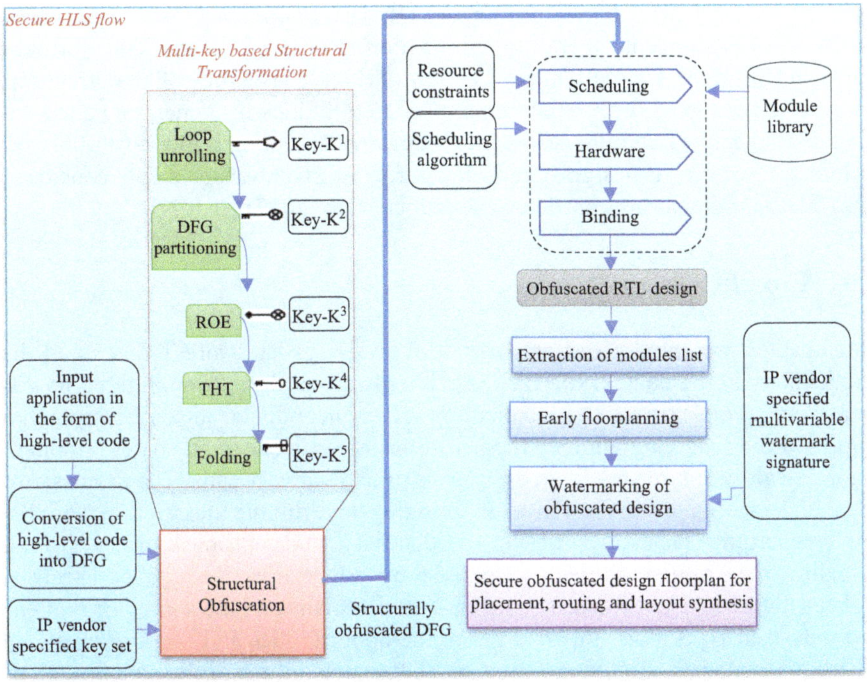

Figure 1.8 Multi-key-based structural obfuscation during HLS

DFG of an input DSP application during behavioral/high-level synthesis. In order to do so, five distinct keys (k1–k5) are used to regulate different obfuscation techniques, such as partitioning, folding knob transformations, loop unrolling, redundant operation elimination, and tree-height transformation. This obfuscates the design at the RTL/gate level without affecting functionality. Next, the security of obfuscated DSP design is further enhanced by watermarking during early floorplanning, which serves as the second line of defense. This acts as a detective control against pirated designs (Sengupta *et al.*, 2019).

Additionally, the approach (Sengupta *et al.*, 2017) is capable of hiding the functionality of the IP design using a series of intertwined HLT techniques employed during HLS. Due to the inter-chaining of multiple algorithmic transformations, the RTL/netlist design of the IP becomes difficult to reverse engineer by an adversary. The primary inputs of this methodology include the CDFG of the application, PSO control parameters (acceleration coefficient, inertia weight, swarm size, etc.), and module library, while the primary output is the structurally obfuscated low-cost RTL IP design. This approach comprises two major blocks, viz., (a) a structurally transformed design based on algorithmic transformation techniques and (b) the PSO-DSE block. The approach produces transformed designs (which are functionally equivalent but structurally different in terms of interconnections/architecture and component density) based on different HLT techniques, viz., (a) Loop Unrolling (LU), (b) Loop Invariant Code Motion (LICM), (c) Redundant Operation Elimination (ROE), (d) Logic Transformation (LT), and (e) Tree Height Transformation (THT). The output of the structurally transformed design is further fed into the PSO-DSE block for generating a low-cost design architecture. The PSO-DSE block explores an optimal, low-cost obfuscated architecture of the IP design based on a series of evolutionary steps, considering local best and global best particles (Sengupta *et al.*, 2017).

1.9 Conclusion

This chapter presented a comprehensive overview of hardware Trojans and their implications in the context of HLS-based hardware design. It highlighted how the increasing complexity and globalization of semiconductor manufacturing have amplified the risks of malicious modifications at various design stages, emphasizing the stealthy nature and diverse payloads of hardware Trojans. The discussion on HLS underscored its pivotal role in bridging algorithmic description and RTL implementation, enhancing design productivity and optimization, while also revealing new vulnerabilities exploitable by adversaries through compromised tools or libraries. The taxonomy of HLS-based Trojans and the survey of academic and industrial HLS tools illustrated both the opportunities and security challenges inherent in modern design automation. Ultimately, the chapter reinforced that ensuring trust in hardware systems demands proactive integration of security measures within HLS flows, promoting the development of resilient, trustworthy, and secure-by-design integrated circuits capable of withstanding emerging threats in the global semiconductor ecosystem.

1.10 Questions and exercises

1. What is a hardware Trojan, and how is it different from a software Trojan?
2. Explain the threat of hardware Trojan attacks in the context of the global design supply chain process.
3. Draw the hardware design lifecycle and discuss the possibility of attacks in different stages.
4. Draw and discuss the classification tree of hardware Trojans.
5. Explain the taxonomy of HLS hardware Trojans in detail.
6. Discuss the high-level synthesis process and its key steps in detail.
7. What are the advantages of the HLS framework?
8. Briefly explain the design space exploration process of HLS.
9. Mention some of the academic and industrial HLS tools and their primary differences.
10. Explain the necessity of Trojan detection and resistance approaches in the context of HLS-generated hardware IP designs.

References

R. S. Chakraborty, S. Narasimhan, and S. Bhunia, (2009), "Hardware Trojan: threats and emerging solutions," *2009 IEEE International High Level Design Validation and Test Workshop*, San Francisco, CA, USA, 2009, pp. 166–171.

H. L. Q. Liu, and J. Zhang, (2016), "A survey of hardware Trojan threat and defense," *Integration*, vol. 55, pp. 426–437.

A. Anshul, and A. Sengupta, (2023a), "A survey of high level synthesis based hardware security approaches for reusable IP cores," *IEEE Circuits and Systems Magazine*, vol. 23, no. 4, pp. 44–62, Fourthquarter 2023.

M. Rostami, F. Koushanfar, and R. Karri, (2014), "A primer on hardware security: models, methods, and metrics," *Proceedings of the IEEE*, vol. 102, no. 8, pp. 1283–1295.

C. Dong, Y. Xu, X. Liu, F. Zhang, G. He, and Y. Chen, (2020), Hardware Trojans in chips: a survey for detection and prevention. *Sensors*, vol. 20, p. 5165.

A. Sengupta, A. Anshul, V. Chourasia, and N. Kumar, (2024), "M-HLS: malevolent high-level synthesis for watermarked hardware IPs," *IEEE Embedded Systems Letters*, vol. 16, no. 4, pp. 497–500.

J. Rajendran, H. Zhang, O. Sinanoglu, and R. Karri, (2013), "High-level synthesis for security and trust," *2013 IEEE 19th International On-Line Testing Symposium (IOLTS)*, Chania, Greece, 2013, pp. 232–233.

C. Pilato, K. Basu, F. Regazzoni, and R. Karri, (2019), "Black-Hat high-level synthesis: myth or reality?," *IEEE Transactions on Very Large Scale Integration (VLSI) Systems*, vol. 27, no. 4, pp. 913–926.

A. Sengupta, (2017a), "Hardware vulnerabilities and their effects on CE devices: design for security against Trojans," *IEEE Consumer Electronics Magazine*, vol. 6, no. 3, pp. 126–133.

A. Sengupta, (2017b), "Hardware security of CE devices," *IEEE Consumer Electronics Magazine*, vol. 6, no. 1, pp. 130–133.

P. Coussy, D. D. Gajski, M. Meredith, and A. Takach, (2009), "An introduction to high-level synthesis," *IEEE Design & Test of Computers*, vol. 26, no. 4, pp. 8–17.

M. Rathor, A. Anshul, and A. Sengupta, (2024), "Securing reusable IP cores using voice biometric based watermark," *IEEE Transactions on Dependable and Secure Computing*, vol. 21, no. 4, pp. 2735–2749, July–Aug. 2024.

A. Anshul, and A. Sengupta, (2023b), "Exploration of optimal crypto-chain signature embedded secure JPEG-CODEC hardware IP during high level synthesis," *Microprocessors and Microsystems*, vol. 102, p. 104916.

A. Sengupta, A. Anshul, and R. Chaurasia, (2023), "Exploration of optimal functional Trojan-resistant hardware intellectual property (IP) core designs during high level synthesis," *Microprocessors and Microsystems*, vol. 103, p. 104973.

C. Pilato, S. Garg, K. Wu, R. Karri, and F. Regazzoni, (2018), "Securing hardware accelerators: a new challenge for high-level synthesis," *IEEE Embedded Systems Letters*, vol. 10, no. 3, pp. 77–80.

W. Hu, C. -H. Chang, A. Sengupta, S. Bhunia, R. Kastner, and H. Li, (2021), "An overview of hardware security and trust: threats, countermeasures, and design tools," *IEEE Transactions on Computer-Aided Design of Integrated Circuits and Systems*, vol. 40, no. 6, pp. 1010–1038.

D. D. Gajski, and L. Ramachandran, (1994), "Introduction to high-level synthesis," *IEEE Design & Test of Computers*, vol. 11, no. 4, pp. 44–54, Winter 1994.

A. Anshul, and A. Sengupta, (2025), "HLS Trojan detection using machine learning technique," *IEEE Embedded Systems Letters*, 2025, doi:10.1109/LES.2025.3614187.

A. Sengupta, S. Bhadauria, and S. P. Mohanty, (2017), "TL-HLS: methodology for low cost hardware Trojan security aware scheduling with optimal loop unrolling factor during high level synthesis," *IEEE Transactions on Computer-Aided Design of Integrated Circuits and Systems*, vol. 36, no. 4, pp. 655–668.

A. Sengupta, and N. Bhui, (2025), "Time-bomb HLS Trojan for performance degradation payload," *IEEE Design & Test*, vol. 42, no. (5), pp. 83–92.

A. Sengupta, A. Anshul, V. Chourasia, and N. Bhui, (2025), "Security vulnerability (Backdoor Trojan) during machine learning accelerator design phases," *IT Professional*, vol. 27, no. 1, pp. 65–72.

M. Abderehman, R. Gupta, R. R. Theegala, and C. Karfa, (2022), "BLAST: belling the Black-Hat high-level synthesis tool," *IEEE Transactions on Computer-Aided Design of Integrated Circuits and Systems*, vol. 41, no. 11, pp. 3661–3672.

M. Rathor, and A. Sengupta, (2024), "Revisiting Black-Hat HLS: a lightweight countermeasure to HLS-aided Trojan attack," *IEEE Embedded Systems Letters*, vol. 16, no. 2, pp. 170–173.

F. Ferrandi, V. G. Castellana, S. Curzel, P. Fezzardi, M. Fiorito, and M. Lattuada (2021), "Invited: Bambu: an open-source research framework for the high-level synthesis of complex applications," *2021 58th ACM/IEEE Design Automation Conference (DAC)*, San Francisco, CA, USA, 2021, pp. 1327–1330.

P. Coussy, C. Chavet, P. Bomel, D. Heller, E. Senn, and E. Martin, (2008), "GAUT: a high-level synthesis tool for DSP applications," in: Coussy, P., Morawiec, A. (eds), *High-Level Synthesis*. Springer, Dordrecht, 2008, pp. 147–169.

A. Canis, *et al.*, (2011), "LegUp: high-level synthesis for FPGA-based processor/accelerator systems," *Proceedings of the 19th ACM/SIGDA International Symposium on Field Programmable Gate Arrays (FPGA '11). Association for Computing Machinery*, New York, NY, USA, pp. 33–36.

S. Gupta, N. Dutt, R. Gupta, and A. Nicolau, (2003), "SPARK: a high-level synthesis framework for applying parallelizing compiler transformations," *16th International Conference on VLSI Design, 2003. Proceedings*, New Delhi, India, 2003, pp. 461–466.

CAD for Assurance, (2025a), Crypto-steganography tool, accessed on Sept. 2025. [Online], *IEEE CEDA and IEEE Hardware Security and Trust Technical*

Committee (HSTTC), Available: https://cadforassurance.org/tools/ip-ic-protection/crypto-steganography-tool/.

CAD for Assurance, (2025b), KHC-Stego tool: key-triggered hash-chaining driven steganography tool, *IEEE CEDA and IEEE Hardware Security and Trust Technical Committee (HSTTC)*, accessed on Sept. 2025. [Online]. Available: https://cadforassurance.org/tools/ip-ic-protection/khc-stego/.

CAD for Assurance, (2025c), Faciometric hardware security tool, accessed on Sept. 2025. [Online], *IEEE CEDA and IEEE Hardware Security and Trust Technical Committee (HSTTC)*, Available: https://cadforassurance.org/tools/ip-ic-protection/faciometric-hardware-security-tool/.

K. Wakabayashi, and B.C. Schafer, (2008), "'All-in-C' behavioral synthesis and verification with CyberWorkBench," in: Coussy, P., Morawiec, A. (eds), *High-Level Synthesis*. Springer, Dordrecht, 2008, pp. 113–127.

Xilinx, (2020), *Vivado Design Suite User Guide: High-Level Synthesis (UG902), Xilinx Inc.*, published 2020 (latest edition).

Cadence Design Systems, (2015), *Stratus high-Level Synthesis Datasheet/User Guide* (Cadence).

Mentor Graphics/Siemens EDA, (2004), *Catapult C Synthesis User Manual/Tool Description* (Mentor/Siemens EDA).

Intel Corporation, (2018), *Intel® High Level Synthesis Compiler Reference Manual* (Intel HLS Compiler). Intel Corporation.

F. Ferrandi, P. L. Lanzi, D. Loiacono, C. Pilato, and D. Sciuto, (2008), "A multi-objective genetic algorithm for design space exploration in high-level synthesis," *2008 IEEE Computer Society Annual Symposium on VLSI*, Montpellier, France, 2008, pp. 417–422.

K. Deb, A. Pratap, S. Agarwal, and T. Meyarivan, (2002), "A fast and elitist multiobjective genetic algorithm: NSGA-II," *IEEE Transactions on Evolutionary Computation*, vol. 6, no. 2, pp. 182–197.

A. Sengupta, and R. Chaurasia, (2024), "Secure FFT IP using C-way partitioning-based obfuscation and fingerprint," *IEEE Design & Test*, vol. 41, no. 5, pp. 55–64.

Y. Lao, and K. K. Parhi, (2015), "Obfuscating DSP circuits via high-level transformations," *IEEE Transactions on Very Large Scale Integration (VLSI) Systems*, vol. 23, no. 5, pp. 819–830.

R. Chaurasia, and A. Sengupta, (2023), "Multi-cut based architectural obfuscation and handprint biometric signature for securing transient fault detectable IP cores during HLS," *Integration, the VLSI Journal*, vol. 95, pp. 102–114.

S. A. Islam, L. K. Sah, and S. Katkoori, (2020), "High-level synthesis of key-obfuscated RTL IP with design lockout and camouflaging," *ACM Transactions on Design Automation of Electronic Systems*, vol. 26, no. 1, Article 6, 2021, 35 pages.

A. Sengupta, and M. Rathor, (2020), "Enhanced security of DSP circuits using multi-key based structural obfuscation and physical-level watermarking for consumer electronics systems," *IEEE Transactions on Consumer Electronics*, vol. 66, no. 2, pp. 163–172.

A. Sengupta, and M. Rathor, (2019), "Protecting DSP kernels using robust hologram-based obfuscation," *IEEE Transactions on Consumer Electronics*, vol. 65, no. 1, pp. 99–108.

C. Pilato, F. Regazzoni, R. Karri, and S. Garg, (2018), "TAO: techniques for algorithm-level obfuscation during high-level synthesis," *2018 55th ACM/ESDA/IEEE Design Automation Conference (DAC)*, SF, USA, pp. 1–6.

A. Sengupta, D. Roy, S. P. Mohanty, and P. Corcoran, (2017), "DSP design protection in CE through algorithmic transformation based structural obfuscation," *IEEE Transactions on Consumer Electronics*, vol. 63, no. 4, pp. 467–476.

A. Sengupta, D. Kachave, and D. Roy, (2019), "Low cost functional obfuscation of reusable IP cores used in CE hardware through robust locking," *IEEE Transactions on Computer-Aided Design of Integrated Circuits and Systems*, vol. 38, no. 4, pp. 604–616.

A. Sengupta, and S. Bhadauria, (2016), "Exploring low cost optimal watermark for reusable IP cores during high level synthesis," *IEEE Access*, vol. 4, pp. 2198–2215.

A. Sengupta, and S. Bhadauria, (2015), "Untrusted third party digital IP cores: power-delay trade-off driven exploration of hardware Trojan secured datapath during high level synthesis," *25th IEEE/ACM Great Lake Symposium on VLSI (GLSVLSI)*, Pittsburgh, Pennsylvania, May 2015, pp. 167–172.

A. Sengupta, S. Bhadauria, and S. P. Mohanty, (2016), "Embedding low cost optimal watermark during high level synthesis for reusable IP core protection," *2016 IEEE International Symposium on Circuits and Systems (ISCAS)*, Montreal, QC, Canada, 2016, pp. 974–977.

A. Sengupta, and M. Rathor, (2021), "Facial biometric for securing hardware accelerators," *IEEE Transactions on Very Large Scale Integration (VLSI) Systems*, vol. 29, no. 1, pp. 112–123.

A. Sengupta, and M. Rathor, (2020), "Securing hardware accelerators for CE systems using biometric fingerprinting," *IEEE Transactions on Very Large Scale Integration (VLSI) Systems*, vol. 28, no. 9, pp. 1979–1992.

Part II

Attack strategies

Chapter 2

Trojan attacks via malicious HLS framework

Anirban Sengupta[1] *and Vishal Chourasia*[1]

High-level synthesis (HLS) has been an industry de facto standard for the past several years. It is a vital cog in the design process of intellectual properties (IP) cores used in the system-on-chip. However, HLS frameworks/tools can be untrustworthy from an IP vendor's/system-on-chip (SoC) integrator's perspective. This is because HLS frameworks/tools are developed by third-party vendors. This chapter discusses the practical possibility of compromising HLS frameworks/tools using malicious HLS Trojan logic. It has been shown in state-of-the-art research that an adversary present in the HLS tool vendor house can compromise the HLS framework for the ulterior motive of compromising IP designs. This chapter provides an elaborate explanation of a recent state-of-the-art malicious HLS framework called "malevolent HLS (M-HLS)" that is capable of incurring performance degradation and denial of service payload due to covertly injected hardware Trojans in the HLS flow. The chapter demonstrates the M-HLS on real-life case studies for highlighting the evasiveness of the Trojan injected (Sengupta *et al.*, 2016; Sengupta and Bhadauria, 2016).

The rest of the chapter is structured as follows: Section 2.1 provides the introduction of the chapter; Section 2.2 discusses the threat model; Section 2.3 provides a brief explanation of "why is HLS framework vulnerable to backdoor hardware Trojan insertion during watermarking?"; Section 2.4 discusses salient features of the chapter; Section 2.5 is based on a detailed discussion of M-HLS; Section 2.6 discusses the Trojan insertion process in the HLS framework; Section 2.7 presents a discussion on the state-of-the-art; and Section 2.8 concludes the chapter by summarizing its significance and implications.

2.1 Introduction

Reusable hardware intellectual properties (IPs) play a crucial role in the design and development of various electronic and multimedia systems. These IPs are specifically designed to validate hardware modules or components, ensuring they meet the required specifications and performance standards. They are designed to cater to a

[1]Deptartment of Computer Science and Engineering, Indian Institute of Technology Indore, India

diverse array of data-intensive applications, from video and image processing and convolution layers to digital signal processing (DSP). By providing a pre-validated, ready-to-deploy solution, these IPs streamline the development process, significantly reducing the time spent on design, testing, and iteration. This leads to faster time-to-market and a more efficient use of resources, as developers can focus on designing complex systems rather than reinventing foundational components. Additionally, the ability to integrate these IPs into various applications without extensive customization further enhances development efficiency, making them a valuable tool for building complex systems in fields such as multimedia, telecommunications, and machine learning. Their reusability makes them highly valuable in system-on-chip (SoC) development, Field-Programmable Gate Array (FPGA) designs, and Application-Specific Integrated Circuit (ASIC) implementations, where efficiency and time-to-market are critical.

The creation of such reusable hardware IPs is commonly facilitated through high-level synthesis (HLS) frameworks. HLS tools enable designers to convert high-level programming languages, like C, C++, or SystemC, into hardware descriptions such as register transfer level (RTL) code, effectively bridging the gap between software development and hardware implementation. These frameworks not only streamline the process of generating efficient hardware designs but also allow for optimization in terms of area, power consumption, and performance, thus enhancing the overall quality of the IPs produced (Anshul and Sengupta, 2023).

While the concept of reusability and efficiency is paramount in modern hardware design, it is crucial to recognize the inherent security vulnerabilities associated with the generation of these IP cores (particularly concerning hardware Trojan attacks by an attacker). HLS offers an efficient way to design complex hardware systems by allowing designers to work with high-level programming languages, abstracting away the intricacies of low-level hardware description languages. While incredibly advantageous for productivity and flexibility, this abstraction creates a potential avenue for malicious actors/attackers to exploit hardware design flow through Trojan attacks. More explicitly, HLS-generated watermarked designs [having the capacity to provide detective countermeasures against IP piracy and false IP ownership claims (Rizzo *et al.*, 2019)] may create security vulnerabilities (in the mux-based interconnect design), which an attacker can exploit to insert backdoor Trojans (Sengupta *et al.*, 2024).

A hardware Trojan is a malicious circuit or logic inserted into an IP design during its design or manufacturing process. These Trojans are typically designed to remain dormant and undetected until triggered by specific conditions. Once activated, they can perform various malicious actions, such as compromising the affected device's performance, security, or privacy. Their ability to remain hidden and dormant and their potential payload make them a significant concern for the cybersecurity community and computer-aided design (CAD) industries. Further, in the competitive landscape of CAD companies, there is a rising distress related to hardware Trojan insertion into IP designs. In the realm of CAD, where innovation and trust form the backbone of every design, these insidious attacks can lead to severe consequences. These Trojans not only compromise the security of designs but also erode the trust between clients and CAD companies. Therefore, research in the direction of HLS-aided Trojans and their proactive countermeasure needs more focus by the security community (Bhunia *et al.*, 2013; Sengupta *et al.*, 2017, 2025).

This chapter discusses a malevolent HLS framework for watermarked hardware IPs that includes different hardware Trojans, which can be automatically inserted in the mux-based interconnect design stage of the HLS flow. This chapter also discusses security vulnerabilities that may exist in HLS-generated watermarked IPs (Sengupta *et al.*, 2024).

2.2 Discussion on the threat model

Designing a compromised HLS tool by an attacker (3rd party HLS tool vendor) aims to hide Trojans during different steps of IP design, such as scheduling, allocation, binding, register allocation, interconnect design, and datapath/controller synthesis. Such HLS frameworks can be called malevolent HLS. A malevolent HLS (M-HLS) tool aims to realize some personal/economic benefits or compromise national-level security. The impact of a malicious HLS tool (containing furtive Trojans) may be data damage (loss/corruption) or parametric degradation (such as performance) or denial of service (open circuit/high impedance/garbage output) (Sengupta *et al.*, 2024). More explicitly, the target of designing a malevolent HLS tool is to create inferior IP components in the market. This can be achieved by exploiting hidden features in the HLS design flow (such as watermarked mux-based interconnect designs, etc.) by inserting malicious backdoor Trojans to generate compromised IPs. Further, compromised HLS tools are part of the national-level attack arsenal, where HLS tools designed in a particular country can be used to develop SoCs in another country (Pilato *et al.*, 2019). Such HLS tools are highly untrustworthy and may be compromised versions (may contain backdoor Trojans without the knowledge of the IP designers). The RTL design produced by an HLS tool is machine-generated and in the order of magnitude bigger and more complex than its high-level specification counterpart. The task of a design engineer becomes highly complex and convoluted while understanding and correlating the generated RTL design with its high-level specification. This makes the process of correlation highly industrious. Moreover, HLS-aided Trojans only get activated under a specific attacker's pre-defined condition. Further, automated test benches/simulations that come along with commercial and academic tools do not provide complete coverage, as exhaustive testing is not pragmatic due to huge time consumption. In addition, some of the HLS-aided Trojans do not produce functional behavioral changes and noticeable side-channel parametric overhead, even on specific triggering conditions. Therefore, they do not become detectable using equivalence checking and side-channel analysis (SCA). Henceforth, simulation-based validation techniques are typically followed as favorable alternatives for validating RTL designs generated through the HLS framework. This provides the underlying motivation for why detecting hardware Trojans at the HLS level is a complex and costly affair (Pilato *et al.*, 2018).

2.3 Why is the HLS framework vulnerable to backdoor hardware Trojan insertion during watermarking?

2.3.1 Trojan vulnerability in HLS-based watermarked IP designs

The watermarked hardware IP designs in the global design supply chain refer to physical representations of hardware design that are secretly marked with a watermark (author information/evidence) to protect against unauthorized use and false IP ownership rights (Sengupta *et al.*, 2018). The primary objective of hardware IP watermarking is to provide a detective countermeasure against IP piracy and false IP ownership claims. An adversary within the design supply chain may pirate the original IP or falsely claim the IP ownership rights for his/her benefit (Sengupta *et al.*, 2018). Therefore, hardware watermarking provides a promising solution against these threats by integrating an IP vendor's secret watermark within the original hardware IP design. The embedded secret mark (i.e., watermark/digital evidence) facilitates a seamless verification during IP piracy detection and validation of the legitimate IP owner in case of an ownership conflict in court (Sengupta and Rathor, 2021). High-level synthesis is a widely used framework for generating watermarked hardware IP designs (Sengupta *et al.*, 2023). HLS is a design methodology that allows hardware designs to be described at a higher level of abstraction, typically using high-level language or a control dataflow graph (CDFG), and then automatically synthesizing into its corresponding register transfer level (RTL) datapath (Koushanfar *et al.*, 2005). In the context of HLS-based hardware IP watermarking, a watermark is a unique signature/identifier embedded into the hardware IP design during the synthesis process. This watermark is typically inserted at a high level of abstraction, such as within the scheduled data flow graph (SDFG) (obtained using its respective CDFG), and is then preserved as the design is transformed into the corresponding RTL datapath. The watermark is embedded as the secret security constraint during the register allocation phase of the HLS process (Sengupta *et al.*, 2018). Post-embedding security constraints, register (color) alterations are performed to satisfy the embedding of security constraints in the respective SDFG (Sengupta *et al.*, 2018). Register allocation information or its corresponding colored interval graph (CIG) is used for embedding security constraints. Subsequently, the register (color) alteration is made to satisfy the security constraints (Sengupta *et al.*, 2018). A final watermark-embedded datapath design is generated and is further converted into its respective synthesizable RTL. Standard approaches (Sengupta *et al.*, 2018; Sengupta and Rathor, 2021; Sengupta *et al.*, 2023; Koushanfar *et al.*, 2005; Gal and Bossuet, 2012; Chen and Schafer, 2021) are some of the established examples of HLS-based watermarking techniques for IP designs. **However, in the process of generating secure IP designs using HLS-based watermarking, certain IP designs expose vulnerabilities that can be exploited by attackers for secret hardware Trojan insertion**. More explicitly, after embedding secret security constraints in the register allocation phase of HLS, the multiplexer-based interconnect design may get altered and yield a free port (input pin of mux) that can be exploited with malicious intent by an attacker to insert a Trojan. Such mis-utilization of free input port(s) in the mux-based interconnect design of registers provides a seamless platform for attackers to realize their malicious intent. Such a possibility of exploitation may also exist in HLS-generated non-watermarked IP designs. We discuss both case studies in our chapter: (a) on watermarked MESA IP core designs and (b) on non-watermarked FIR IP designs. Note: The FIR and MESA benchmarks were adopted from the University of California, Santa Barbara (UCSB, 2023) Express Benchmark. MESA Horner Bezier is a well-known digital media application used in several DSP and multimedia cores of consumer electronics systems (Wolinski *et al.*, 2010). The mux-based interconnect designs showing vulnerabilities that can be exploited by attackers for secret hardware Trojan insertion (corresponding to both watermarked and non-watermarked IP designs) are elaborated in the next subsection.

2.3.2 Examples to highlight the HLS Trojan vulnerability

Figures 2.1 and 2.2 show MESA Horner Bezier's scheduled data flow graph (SDFG) without and with a watermark, respectively (Sengupta *et al.*, 2024; Mohanty *et al.*, 2008; Sengupta and Mohanty, 2019; Sengupta and Roy, 2017). The embedded watermark is adopted from Sengupta and Roy (2017). Scheduling is performed based on resource constraints of four multipliers (M1 to M4) and three adders (A1 to A3), as shown in the figures. In the SDFG, four unique registers (named R, G, B, and Y) are used to execute 18 storage variables, where a distinct color has been used to denote each register. The assignment of all storage variables to the registers in seven different control steps (C0 to C6) is shown in the figures. Similarly, Figure 2.3 shows a scheduled data flow graph of an FIR without a watermark. Scheduling is performed based on resource constraints of two multipliers (M1 to M2) and two adders (A1 to A2), as shown in Figure 2.3. In the SDFG,

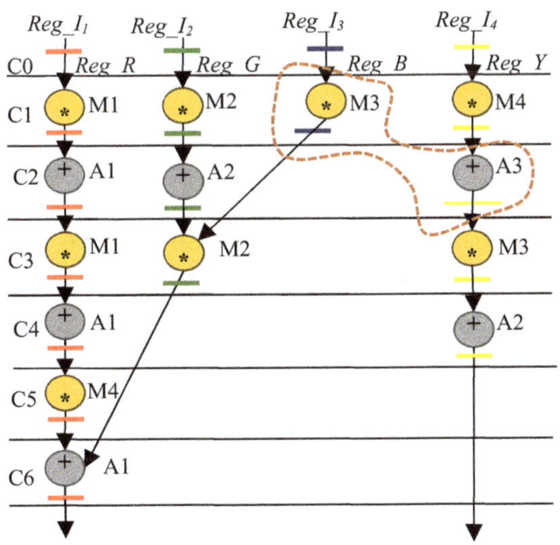

Figure 2.1 Scheduled data flow graph of MESA Horner Bezier without watermark

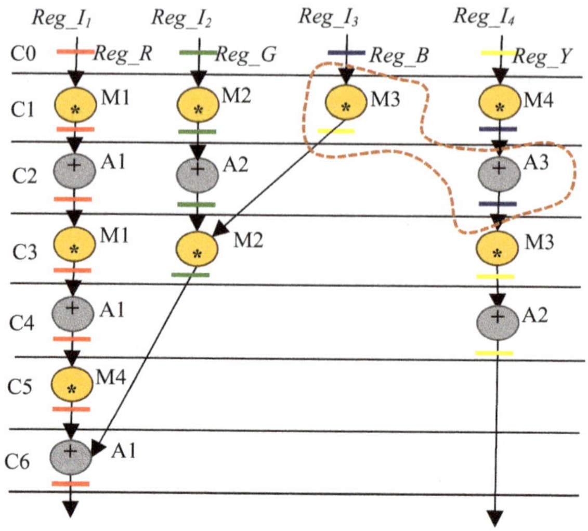

Figure 2.2 *Scheduled data flow graph of MESA Horner Bezier with embedded watermark*

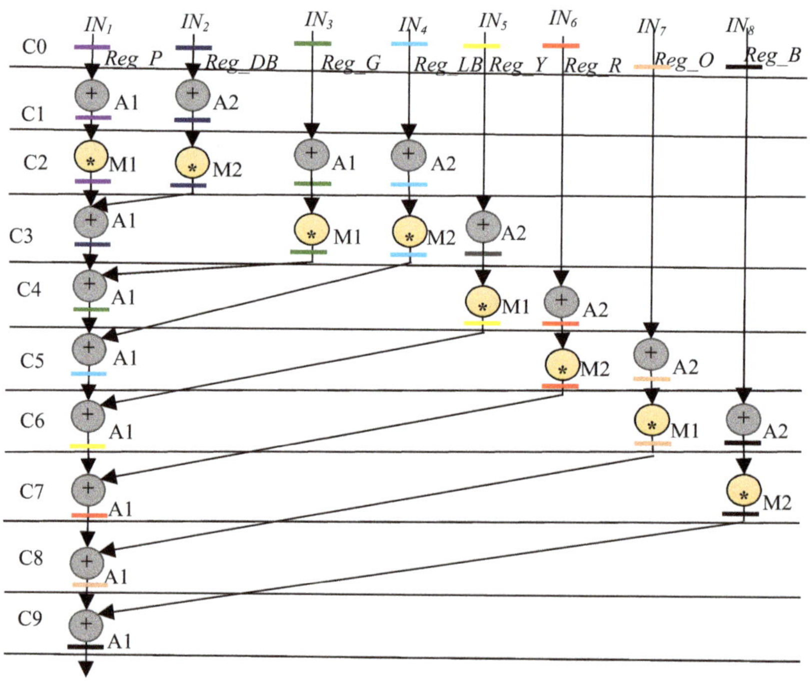

Figure 2.3 *Scheduled data flow graph of FIR without watermark*

eight unique registers (named P, DB, G, L, Y, R, O, and B) are used to execute 31 storage variables, where a distinct color has been used to denote each register. The assignment of all storage variables to the registers in ten different control steps (C0 to C9) is shown in Figure 2.3. Subsequently, Figures 2.4 and 2.5 depict the mux-based interconnect design information of the data path without and with a watermark in **Register B,** corresponding to Figures 2.1 and 2.2, respectively. The dotted red color lines indicate the portion of the register allocation affected due to the embedded watermark. As observed in Figure 2.5, one input/output port (shown in orange color) in the mux-interconnect design of **Register B** post-watermarking is unutilized (free), creating a vulnerability that an attacker can exploit for secret hardware Trojan insertion. Earlier, there were no free input/output ports, as shown in Figure 2.4 (all pins of the 2×1 mux corresponding to **Register B** are occupied). **Therefore, hardware watermarking can create a potential vulnerability in the final hardware IP design for secret hardware Trojan insertion.** Further, as discussed above, such cases can also exist in HLS-generated non-watermarked IP designs. Figure 2.3 shows the SDFG of FIR without watermarking, and Figure 2.6 shows its respective mux-based interconnect design of the data path without a

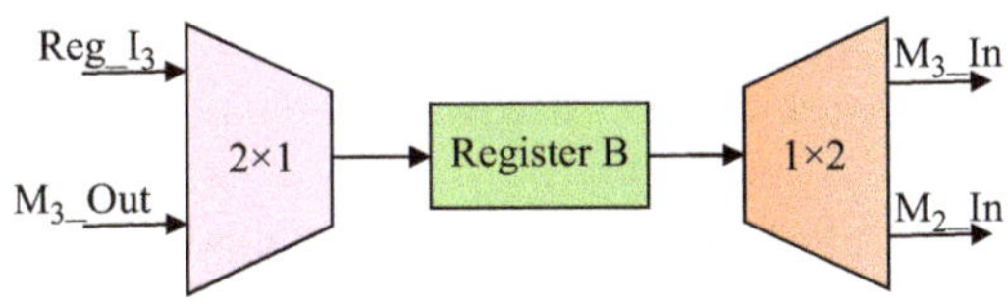

*Figure 2.4 Mux-based interconnect design of datapath **without watermark** in Register B corresponding to Figure 2.1*

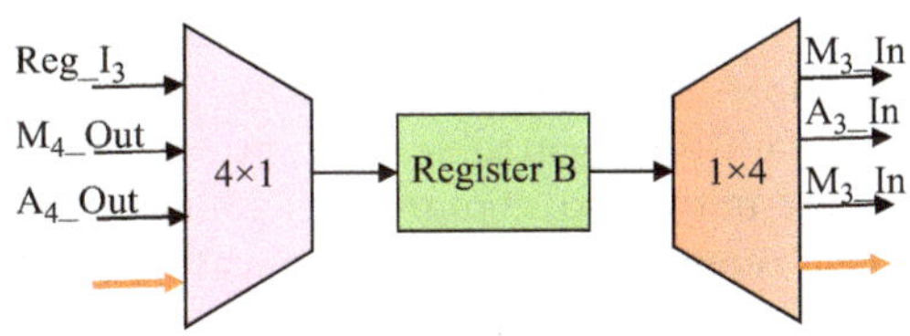

*Figure 2.5 Mux-based interconnect design of datapath **with watermark** in Register B corresponding to Figure 2.2*

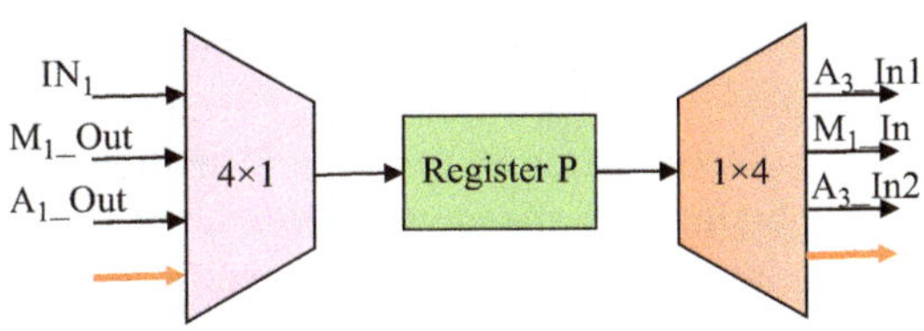

Figure 2.6 Mux-based interconnect design of datapath without watermark in Register P corresponding to Figure 2.3

watermark for **Register P**. As depicted in Figure 2.6, one input/output port (shown in orange color) in the mux-interconnect design of **Register P** is unutilized. An attacker can misuse these free input/output ports to insert a hardware Trojan in the HLS-generated non-watermarked IP designs.

2.4 Salient features of the chapter

- This chapter elaborates a state-of-the-art malevolent HLS framework that explains the possibility of hardware Trojan insertion by exploiting vulnerability in watermarked IP designs.
- Highlights three different HLS-aided hardware Trojans that can be secretly inserted by an attacker in the mux-based interconnect design stage of the HLS flow, viz., performance degradation hardware Trojan (PD-HT) and denial of service (DoS) hardware Trojan (DoS-HT). These Trojans get activated under a specific attacker's pre-defined condition.
- Demonstrate algorithms to automatically insert the hardware Trojans in the HLS flow. More explicitly, this chapter discusses the following algorithms:
    a) An algorithm to insert a performance degradation hardware Trojan (PD-HT) in the IP design during HLS.
    b) An algorithm to insert a DoS hardware Trojan (DoS-HT) in the IP design during HLS.
    c) An algorithm to generate the final RTL datapath corresponding to a hardware Trojan-infected IP design.

2.5 Detailed discussion on Malevolent HLS

Figure 2.7 illustrates the framework for malevolent high-level synthesis, showcasing its overall structure and key components (Sengupta *et al.*, 2024). The M-HLS framework is adaptable/compatible with the classical/standard HLS flow. The primary inputs of the framework comprise (a) a high-level description/transfer function or intermediate representation of the application (in the form of CDFG), (b) IP-vendor chosen resource configuration (can be obtained using design space exploration), (c) IP-vendor selected watermark information, and (d) the module library corresponding to different functional components (such as adder, comparator, multiplier, etc., provided by the HLS vendor). The HLS framework accepts these primary inputs and processes the input application. In HLS, scheduling, allocation, and binding are the key steps that transform an abstract algorithmic representation into its respective hardware form (Sengupta *et al.*, 2024, 2018). Scheduling involves the determination of timing information for each operation in the CDFG. SDFG is obtained as the output post-scheduling. Next, allocation involves mapping operations from the CDFG to specific hardware resources like functional units (adders, multipliers, etc.), registers, and memory. Binding is the final step, where operations are bound to specific resources or components within the hardware (Sengupta *et al.*, 2018). Subsequently, the watermark (signature) is

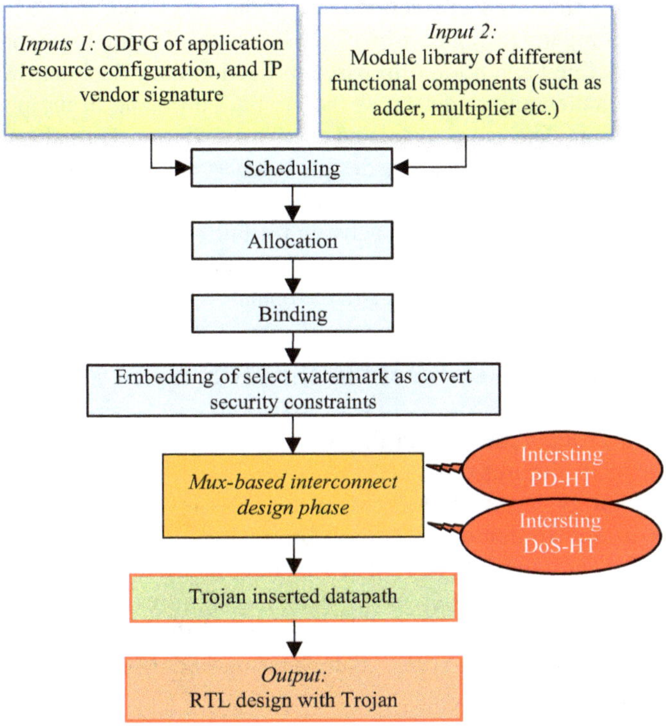

Figure 2.7 Malevolent high-level synthesis framework

embedded in the form of security constraints during the register allocation phase of HLS, followed by corresponding register alterations to satisfy the security constraints (discussed in Section 2.3). Further, the mux-based interconnect design information is generated using the corresponding multiplexing scheme. As discussed in Section 2.3, watermarked HLS-generated IP designs expose vulnerabilities that can be exploited by attackers for secret hardware Trojan insertion. The target of an attacker is to insert the Trojan in a safe location (such that it goes unnoticed during normal validation/testing/inspection) by leveraging the unutilized ports in the mux-interconnect design information of registers.

This chapter discusses two different hardware Trojans that an attacker can secretly insert by utilizing the existing hardware vulnerability. The first type of Trojan is the performance degradation hardware Trojan (PD-HT). A PD-HT is a malicious modification or insertion of a hardware component within an integrated circuit (IC) or IP design file. Its primary purpose is not to actively disrupt the function of the system but to enable performance degradation on specific triggering conditions. The second type of Trojan consists of denial of service (DoS) hardware Trojans (DoS-HT). The primary goal of DoS-HT is to induce denial of service in the form of an open circuit condition/high impedance state/indeterministic state on a specific triggering condition. The inserted hardware Trojans are activated only under a specific attacker's pre-defined triggering condition. The trigger is controlled using comparator logic, which activates the Trojan logic on a specific attacker's pre-defined condition (discussed in Section 2.6). Such hardware Trojans are extremely challenging to detect, as they remain dormant during normal operation and execution. Therefore, detecting the Trojan during normal validation and testing is extremely tedious (as the design behaves normally without functional/timing change under non-triggering conditions, i.e., the inserted Trojan remains dormant until the attacker wishes to activate it). Further, the PD-HT logic, even on specific conditional activation, only affects the performance (not the function); therefore, it is even harder to detect such Trojans.

2.6 Trojan insertion process in the HLS framework

The hardware Trojans inserted in the mux-based interconnect design information are shown in Figures 2.8–2.12. The mux-interconnect design details corresponding to Register P of the FIR IP and Register B of the watermarked MESA IP are used to demonstrate the Trojan designs inserted. Trojan designs can be inserted into any HLS-based IPs as long as there is an unutilized input port of the mux of the register in the datapath. Figures 2.8 and 2.9 show the insertion of PD-HT in the unutilized free port of the mux of Register P and Register B corresponding to FIR and watermarked MESA IP, respectively.

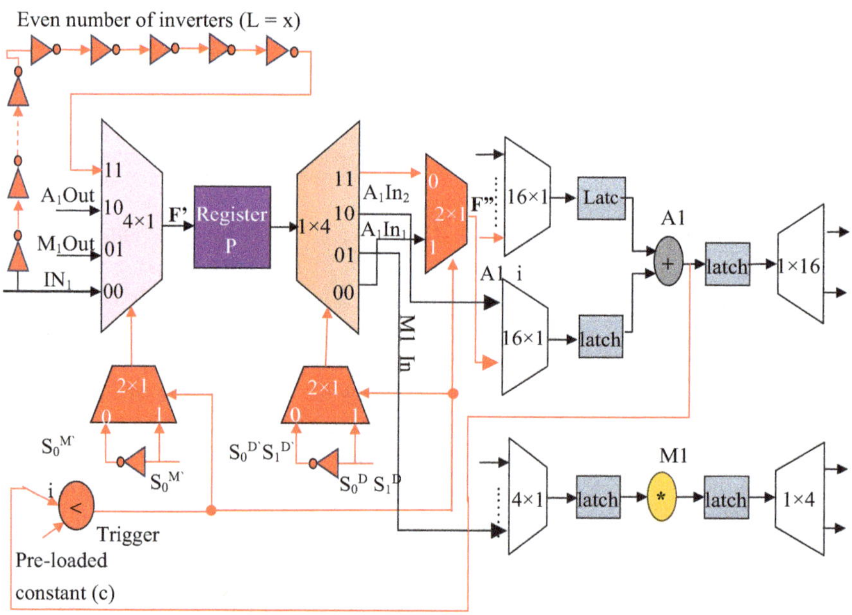

Figure 2.8 Partial datapath of an FIR IP depicting insertion of PD-HT in the multiplexer of ***Register P***

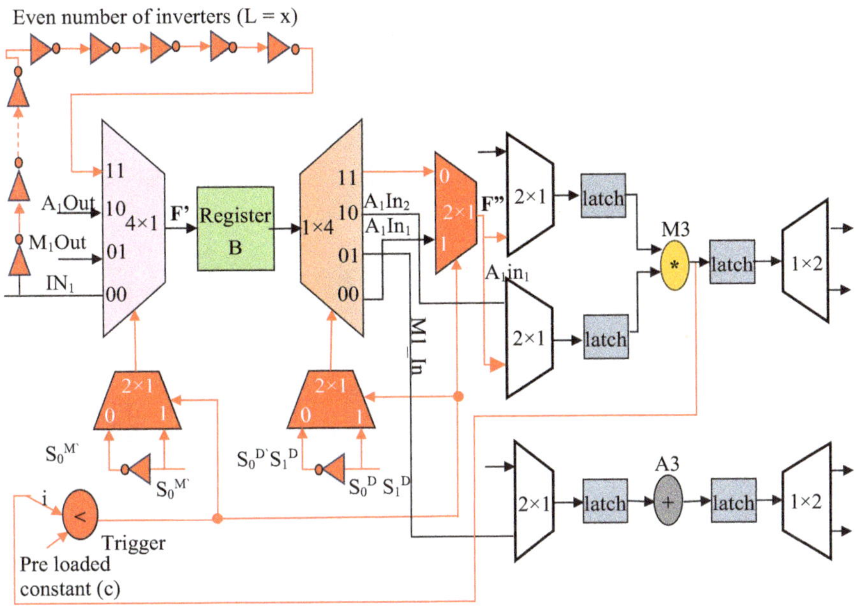

Figure 2.9 Partial datapath of a watermarked MESA IP depicting insertion of PD-HT in the multiplexer of ***Register B***

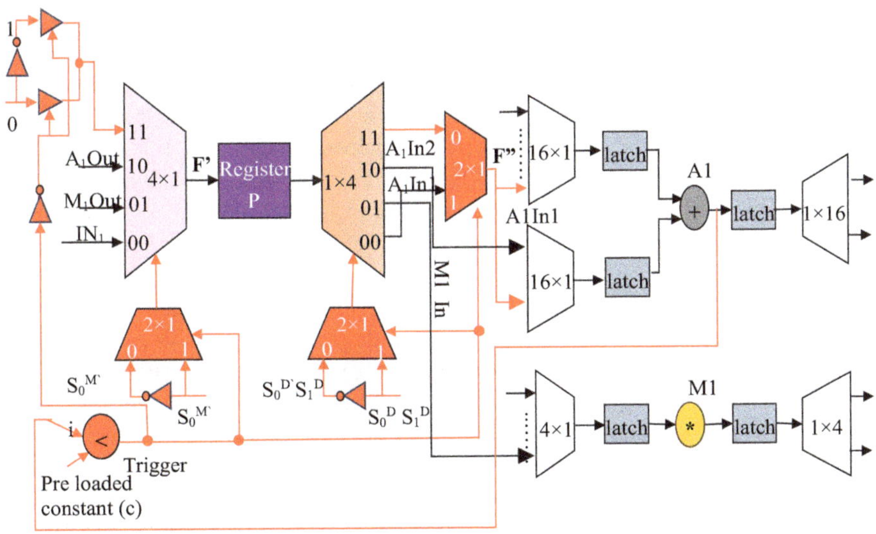

Figure 2.10 Figure depicting the insertion of DoS-HT in the multiplexer of ***Register P*** *in the FIR IP*

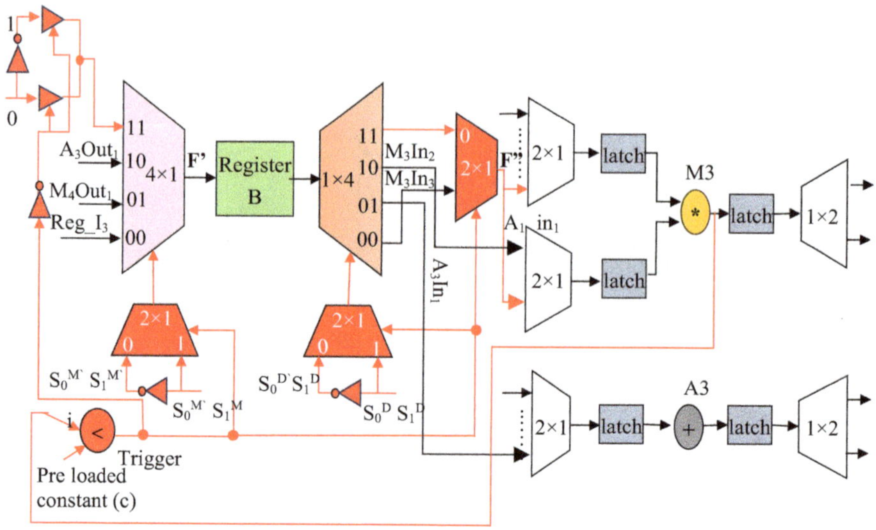

*Figure 2.11 Partial datapath of a watermarked MESA IP depicting insertion of DoS-HT in multiplexer **Register B***

A performance degradation hardware Trojan is a malicious circuit modification that, when triggered, causes reduced performance (larger delays) in an IP design without altering its functionality. As shown in Figures 2.8 and 2.9, the trigger is connected with the output of a comparator. When the value of "*I*" (one of the comparator inputs) becomes equal to the attacker's pre-defined loaded constant "*c*," the Trojan circuit gets activated. The input "*i*" is connected internally with the output of the adder (*A1*) in the partial datapath of a FIR IP (as shown in Figure 2.8). Similarly, the input "*I*" is connected internally with the output of the multiplier (M3) in the partial datapath of a watermarked MESA IP (as shown in Figure 2.9). Further, Table 2.1 shows the behavioral table and output corresponding to the mux of Register P (as shown in Figure 2.8) highlighted in the Trojan active condition. When $i = c$ (i.e., trigger condition), then F" = delayed output of IN1 (which leads to performance degradation). However, Table 2.2 shows the behavioral table and output corresponding to the mux of Register P (as shown in Figure 2.8) highlighted in the Trojan inactive condition at $i \neq c$, F" = IN1(A_1In_1), which shows the circuit's intended normal output without delay. As depicted in Figures 2.8 and 2.9, the attacker's inserted even number (*L*) of inverters causes performance degradation by causing propagation delay. Moreover, the attacker can vary the performance degradation by controlling the number of inverters secretly inserted in the design.

The complete analysis regarding performance degradation variation is discussed in the analysis section. As discussed above, the triggering conditions for all Trojans are the same.

Pseudocode 1: Algorithm to insert performance degradation hardware Trojan (*PD-HT*) in the IP design during HLS.

Input: Register allocation information of IP and its corresponding mux-based interconnect design information.
Output: Trojan-infected datapath.

FOR each register (r) in R
  IF ($\phi_{in} \in \psi$)
    1. INSERT(Δ_1, ψ) // Insert Δ_1 in ψ
      Output (Δ_1) ← Select line of the Register_Mux containing ϕ_{in} in ψ
    2. INSERT(Δ_2, ψ) // Insert Δ_2 in ψ
      Output (Δ_2) ← Select line of the Register_DeMux containing ϕ_{out} in ψ
    3. INSERT(Δ_3, ψ) // Insert Δ_3 in ψ
        Input$_1$(Δ_3) ← ϕ_{out} of selected DeMux
      Input$_2$(Δ_3) ← First output of selected DeMux
      Output (Δ_3) ← Input of respective Mux of functional unit (FU, *i.e.*,adder/multiplier) in ψ
    4. INSERT(δ, ψ) // Insert δ in ψ
      Input(δ) ← First input of selected mux containing ϕ_{in} in ψ
        Output(δ) ← ϕ_{in} port of the selected mux
  END IF

END FOR

INSERT(σ, ψ) // *insert* σ in ψ

Input$_1$(σ) ← c

Input$_2$(σ) ← the output of the respective functional unit //the FU having the output of Δ_3 as one of its input

Output(σ) ← Select lines of all inserted Δ_1, Δ_2, *and* Δ_3 in ψ

$\psi' = f(\textit{Registers, Register\_Muxes, Register\_DeMuxes}, \Delta_1, \Delta_2, \Delta_3, c, \sigma)$

Figure 2.12 Algorithm to insert performance degradation hardware Trojan (PD-HT) in the IP design during HLS

Table 2.1 Behavioral table and output corresponding to Mux of ***Register P*** *in Figure 2.8 (PD-HT)—highlighting Trojan active condition*

Trigger condition	Comparator o/p (Q)	S_0^M	S_1^M	F'	Output
$i = c$ (Trojan active)	0	0	0	IN1	Delayed output

Table 2.2 Behavioral table and output corresponding to Mux of Register P in Figure 2.8 (PD-HT)—highlighting Trojan inactive condition

Trigger condition	Comparator o/p (Q)	S_0^M	S_1^M	F'	Output
$i \neq c$ (Trojan inactive)	1	0	0	IN1	Normal
	1	0	1	M_1_Out	
	1	1	0	A_1_Out	

Table 2.3 Behavioral table and output corresponding to Mux of Register P in Figure 2.10 (DoS-HT)—highlighting Trojan active condition

Trigger condition	Comparator o/p (Q)	S_0^M	S_1^M	F'	Output
$i = c$ (Trojan active)	0	0	0	Z	Z (DoS)

Table 2.4 Behavioral table and output corresponding to Mux of Register P in Figure 2.10 (DoS-HT)—highlighting Trojan inactive condition

Trigger condition	Comparator o/p (Q)	S_0^M	S_1^M	F'	Output
$i \neq c$ (Trojan inactive)	1	0	0	IN1	Normal
	1	0	1	M_1_Out	
	1	1	0	A_1_Out	

Subsequently, Figures 2.10 and 2.11 show DoS-HT insertion in the unutilized free port of the mux of Register P and Register B corresponding to FIR IP and watermarked MESA IP, respectively. A denial of service (DoS) hardware Trojan is a concealed circuit modification that, upon activation, prevents a specific component or function within a digital system from operating, rendering it non-operational. As discussed above, the triggering condition is the same as PD-HT triggering (shown in Figures 2.10 and 2.11). Further, Table 2.3 shows the behavioral table and output corresponding to the mux of Register P (shown in Figure 2.10) highlighted in the Trojan active condition. When $i = c$ (i.e., trigger condition), then F" = Z (which leads to an undeterministic state resulting in a denial of service from the multiplexer output). However, Table 2.4 shows the behavioral table and output corresponding to the mux of Register P (shown in Figure 2.11) highlighted in the Trojan inactive condition at $i \neq c$, F" = IN1(A_1In_1) produces the circuit's intended normal behavior. An attacker preloads the pre-defined constant "*c*" in a register during the Trojan insertion process. The algorithmic details (in the form of pseudocode) of the automatic insertion of the Trojans in the IP datapath generated through HLS are discussed below.

The description of symbols/notations used in the pseudocodes corresponding to HLS-aided Trojans is explained in the initial description box. Subsequently, pseudocode 1 (Figure 2.12) depicts the algorithm to insert PD-HT in the IP design during HLS. Similarly, pseudocode 2 (Figure 2.13) reports the algorithm to insert DoS-HT in the IP design during HLS. Finally, pseudocode **3** (Figure 2.14) depicts the algorithm to generate a Trojan-inserted IP datapath design. All input, output, and used symbols are explained in the initial description box. The malevolent (Trojan-inserted) designs have been implemented by Sengupta *et al.* (2024) in the

Pseudocode 3: Algorithm to insert DoS hardware Trojan (*DoS-HT*) in the IP design during HLS

Input: Register allocation information of IP and its corresponding mux-based interconnect design information.

Output: Trojan-infected datapath.

```
FOR each register (r) in R
    IF (ϕ_in ∈ ψ)
        1. INSERT(Δ1, ψ)        // Insert Δ1 in ψ
           Output (Δ1) ← Select line of the Register_Mux containing ϕ_in in ψ
        2. INSERT(Δ2, ψ)        // Insert Δ2 in ψ
           Output (Δ2) ← Select line of the Register_DeMux containing ϕ_out in ψ
        3. INSERT(Δ3, ψ)        // Insert Δ3 in ψ
           Input1(Δ3) ← ϕ_out of selected DeMux
           Input2(Δ3) ← First output of selected DeMux
           Output (Δ3) ← Input of respective Mux of functional unit
           (FU, i.e., adder/multiplier) in ψ
        4. INSERT((δ', ψ), (λ1, ψ), (λ2, ψ)) // Insert δ', λ1, λ2 in ψ
           Input(δ') ← Output(σ)
           Input(λ1) ← 0
           Input(λ2) ← NOT Input(λ1)
           Output(δ') ← the enable inputs ( en) of λ1 and λ2
           Output(λ1 and λ2) ← ϕ_in of selected Mux
    END IF
END FOR
INSERT(σ, ψ)        // insert σ in ψ
Input1(σ) ← c
Input2(σ) ← the output of the respective functional unit   //the FU having the output of Δ3 as one of its input
Output(σ) ← Select lines of all inserted Δ1, Δ2, and Δ3 in ψ
ψ' = f(Registers, Register_Muxes, Register_DeMuxes, Δ1, Δ2, Δ3, c, σ )
```

Figure 2.13 Algorithm to insert DoS hardware Trojan (DoS-HT) in the IP design during HLS

Pseudocode 4: Algorithm to generate final RTL datapath corresponding to hardware Trojan-infected IP design

Input: Different Hardware Trojan-infected interconnect design.

Output: Final hardware Trojan-infected RTL datapath corresponding to IP design.

```
S ← Set of Trojan-infected interconnect designs
FOR each s in S
        Accept s as input and generate its corresponding RTL
        datapath of IP using HLS
END FOR
```

Figure 2.14 Algorithm to generate the final RTL datapath corresponding to a hardware Trojan-infected IP design

publicly available watermarking HLS framework/tool (Crypto-Steganography Tool, 2023; Sengupta *et al.*, 2018). The watermarking HLS framework/tool (Crypto-Steganography Tool, 2023; Sengupta *et al.*, 2018) is capable of generating watermarked IPs using a key-based crypto-framework. However, the Trojans can be inserted in any HLS framework/tool that has the capability to generate watermarked IP designs, such as the CAD for Assurance KHC-Stego Tool (2023), the CAD for Assurance Faciometric Hardware Security Tool (2023) (Koushanfar *et al.*, 2005; Gal and Bossuet, 2012; Chen and Schafer, 2021).

2.7 Discussion on state-of-the-art

There exist few works in the state-of-the-art on HLS-aided hardware Trojan, such as Sengupta *et al.* (2017), Pilato *et al.* (2019), Sengupta *et al.* (2015), and Rajendran *et al.* (2013). Approach (Pilato *et al.*, 2019), performed by injecting different types of hardware Trojans in the HLS flow by showcasing its functional and non-functional effects. Pilato *et al.* (2019) reveal that malicious use of HLS tools can alter circuits by adding latency, depleting energy, or compromising the security of cryptographic hardware cores. However, Pilato *et al.* (2019) show the insertion of Trojans only in normal, i.e., HLS-generated, non-watermarked IP designs. In contrast to the M-HLS framework (Sengupta *et al.*, 2024), Pilato *et al.* (2019) do not discuss the Trojan insertion by exploiting the security vulnerability of HLS-generated watermarked IPs. Further, the Trojans discussed in Pilato *et al.* (2019) depict a larger area and power overhead as well as achieve lesser performance degradation than the hardware Trojans discussed in Sengupta *et al.* (2024) (which is explained in Section 2.8). Pilato *et al.* (2019) discussed the insertion of Trojans in HLS-generated finite state machines (FSMs) for achieving performance degradation, the insertion of Trojans in scheduling/binding for battery exhaustion attacks, and the alternative insertion of Trojans in the datapath for downgrade attacks under the attacker's triggering conditions. However, the M-HLS framework (Sengupta *et al.*, 2024) inserts PD-HT and DoS-HT in the mux-based interconnect design stage of HLS flow to achieve performance degradation and denial of service (non-operational) under the attacker's specific rare event trigger activation. This chapter highlights hardware Trojan insertion by exploiting the security vulnerability present in the mux-based interconnect design of registers in HLS-generated watermarked IP designs. The M-HLS Trojan insertion framework also applies to HLS-generated non-watermarked IP designs. Moreover, the HLS-aided Trojan insertion framework is capable of achieving larger performance degradation from an attacker's perspective than Pilato *et al.* (2019). Further, the approach (Sengupta *et al.*, 2017) has discussed functional Trojans in the micro third-party intellectual property (3PIP) of the HLS tool library. However, Sengupta *et al.* (2017) do not highlight performance degradation-based Trojan and denial of service-based Trojan, viz., PD-HT and DoS-HT, unlike the M-HLS approach. Additionally, Sengupta *et al.* (2017) do not exploit watermarked IPs to covertly insert Trojans in the interconnect design phase of HLS, unlike the M-HLS approach. Furthermore, approaches by Sengupta and Bhadauria (2015) and Rajendran *et al.* (2013) discuss functional Trojans in the HLS library of CAD tools. However, both these approaches present Trojans that are functional by nature and do not affect performance, data integrity, and service, unlike the PD-HT.

Additionally, some hardware Trojan detection techniques are available in state-of-the-art. Some of the Trojan detection mechanisms include equivalence checking (Abderehman *et al.*, 2022), DMR-based security-aware scheduling (Sengupta *et al.*, 2017), automatically generated hardware test benches (Mondal *et al.*, 2023), side-channel analysis (Tehranipoor and Koushanfar, 2010), functional verification by test vector (Wang *et al.*, 2008), detection using physical inspection and reverse engineering (Ludwig *et al.*, 2021), detection using manufacturing fault test (Mondal *et al.*, 2019), fault-secured HLS technique (Wu and Karri, 2004), and switching probability analysis or transition probability-based method (Nejadmoghadam *et al.*, 2014; Mohapatra *et al.*, 2019). It has already been discussed in the literature earlier (Sengupta *et al.*, 2017; Pilato *et al.*, 2019; Rajendran *et al*, 2013) that HLS-aided Trojans are extremely stealthy by nature, therefore extremely challenging to detect. More explicitly, it has been established and proved earlier in Sengupta *et al.* (2017), Pilato *et al.* (2019), and Rajendran *et al* (2013) that the detection techniques of Sengupta *et al.* (2017), Bushnell and Agrawal (2013), Mondal *et al.* (2023), Tehranipoor and Koushanfar (2010), Wang *et al.* (2008), Ludwig *et al.* (2021), Mondal *et al.* (2019), Wu and Karri, (2004), Nejadmoghadam *et al.* (2014), and Mohapatra *et al.* (2019) are not adequately equipped to detect HLS-aided Trojans. Therefore, the HLS-aided Trojans (which are non-functional by nature and are only triggered under a single pre-defined condition) cannot be detected by state-of-the-art detection schemes. The following paragraph discusses the limitations of these detection techniques and why they are not capable of PD-HT and DoS-HT.

As discussed in Section 2.2, the RTL design generated by an HLS tool is significantly larger and more intricate than its high-level specification counterpart. Understanding and correlating this complex RTL design with the original specification is tedious/time-consuming and poses a daunting challenge for the designer. This has also been discussed in Pilato *et al.* (2019). Additionally, HLS-aided DoS-HT activates only under a specific pre-defined condition set by the attacker. The Trojan-infected hardware IP will behave functionally the same under all conditions except for the triggering condition. Moreover, the PD-HT only impacts performance degradation without affecting the functionality and/or data. Equivalence checking (Abderehman *et al.*, 2022) considers the availability of Trojan-infected RTL code (generated by exploiting the Bambu tool), which is an important file used for launching equivalence checking analysis. However, the Bambu tool is not capable of generating Trojan-infected RTL codes for watermarked IP designs. Therefore, the equivalence checking technique is not capable of detecting HLS PD-HT (Sengupta *et al.*, 2024). Moreover, equivalence checking (Abderehman *et al.*, 2022) only works for BE-HT and DA-HT attacks presented in (Pilato *et al.*, 2019). Hence, the equivalence checking technique does not work for DoS-HT (Sengupta *et al.*, 2024). Side-channel analysis techniques, such as those used by Huang *et al.* (2018), are not equipped to detect PD-HT and DoS-HT kind HLS Trojans. This is because both PD-HT and DoS-HT do not leak significant side-channel information (such as power, energy, timing, etc.). The detection technique (Rathor and Sengupta, 2024) has been proven to effectively detect the Black-HAT HLS Trojan, viz., BE-HT. However, it is not capable of detecting PD-HT and DoS-HT. The detection technique (Yasaei *et al.* 2022) has assumed a threat model where a rogue in-house designer or third-party EDA tool vendor directly inserts HT in the Verilog netlist of the RTL IP, which is very different from the threat model considered in Sengupta *et al.* (2024). This is because Sengupta *et al.* (2024) consider HT insertion in the pre-RTL synthesis stage. Further, Yasaei *et al.* (2022) are only compliant with Verilog code (not VHDL). It uses the Pyverilog toolkit for feeding into the graph neural network (GNN) model. It is not capable of handling watermarked VHDL codes used in Sengupta *et al.* (2024) for PD-HT and DoS-HT HLS Trojan kinds. Furthermore, the Trojan-infected RTL code is significantly harder and more cumbersome to detect through physical inspection and reverse engineering (Ludwig *et al.*, 2021), given the time-to-market and cost constraints. Additionally, Trojans are strategically inserted into specific design parts and can be activated by a single trigger signal, making them difficult to detect through reverse engineering. Moreover, tests designed for detecting manufacturing faults (Mondal *et al.*, 2019), like stuck-at faults and bridging faults, cannot ensure the detection of these Trojans. Further, Sengupta *et al.* (2017) incur larger area/power overhead and dependency on at least two different IP vendors due to the double module redundancy framework for detection. Moreover, automated test benches (Mondal *et al.*, 2023) and simulations provided by commercial and academic tools cannot provide complete coverage because exhaustive testing is impractical and time-consuming. Similarly, the exhaustive testing scenario and dormant nature (due to a specific pre-defined trigger condition) of PD-HT and DoS-HT Trojans make them resilient against Wang *et al.* (2008). Next, the fault-secure HLS addresses resiliency against transient faults only (Wu and Karri, 2004). Switching/transition probability analysis detection methods (Nejadmoghadam *et al.*, 2014; Mohapatra *et al.*, 2019) do not work for Trojans because watermarked IPs of signal processing algorithms (comprising a massive number of gates), evaluating the transition probability of all gates, can be computationally prohibitive and impractical. So far in the literature, such transition probability detection techniques (Nejadmoghadam *et al.*, 2014; Mohapatra *et al.*, 2019) have been demonstrated only on ISCAS benchmarks, which are significantly less complex in terms of gate complexity than HLS IPs.

2.8 Analysis of case studies

Authors in Sengupta *et al.* (2024) have evaluated and analyzed Trojan-infected design using the NanGate 15 nm standard cell library (Open Cell NanGate Library, 2024). The impact of the PD-HT and DoS-HT insertion on design metrics such as area, performance, and power is systematically reported across Figures 2.15–2.20. Figure 2.15 summarizes the area overhead, measured in terms of gate count, that arises when the PD-HT/DoS-HT is inserted into the watermarked IP design. As shown in Figure 2.15, PD-HT and DoS-HT incur, on average, a nominal area overhead *w.r.t.* the baseline design. This indicates that the Trojan remains lightweight while still being stealthy. The area overhead due to the Trojan attack is further analyzed in Figure 2.16, where the attacker-controlled parameter, the number of inserted inverters (denoted as x), is systematically varied to yield performance degradation. Figure 2.16 shows the resulting area overhead achieved for the watermarked PD-HT inserted MESA IP design with variation in the attacker's selected number of inverters (X). Similarly, Figure 2.17 illustrates performance degradation achieved for watermarked MESA IP inserted with HLS-aided PD-HT with variation in the attacker's selected number of inverters (X). A clear trend is noticeable: as the number of inverters increases, the level of performance

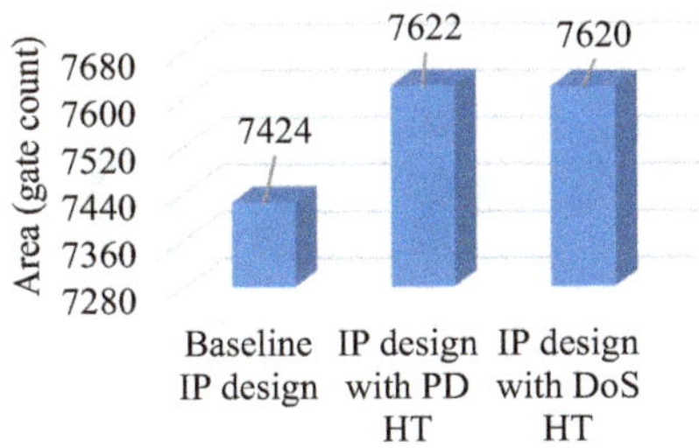

Figure 2.15 Area (gate count) due to insertion of PD-HT (Sengupta et al., 2024) in watermarked MESA IP design

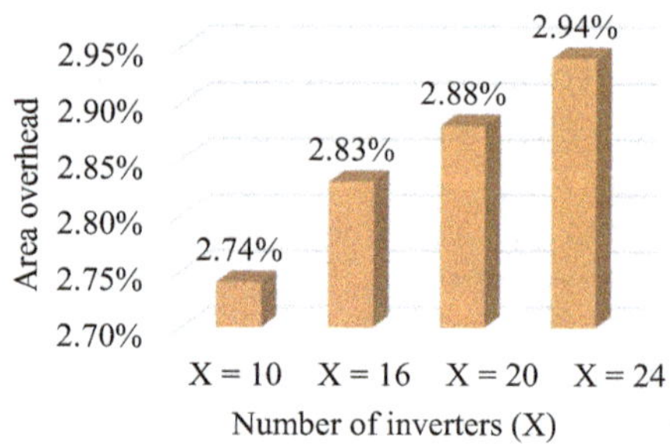

Figure 2.16 Area overhead achieved for watermarked MESA IP inserted with PD-HT (Sengupta et al., 2024) with variation in attacker's selected number of inverters (X)

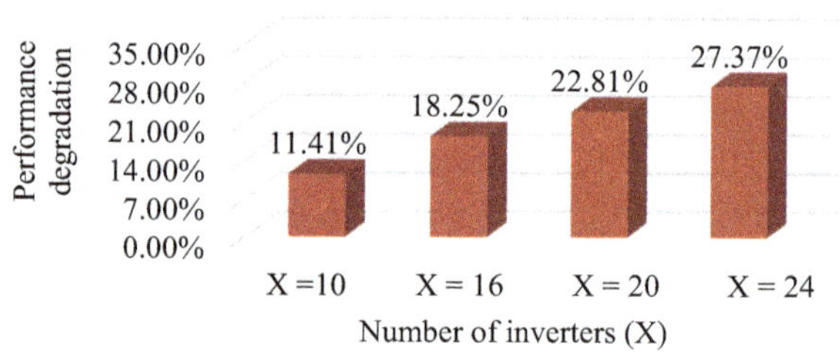

Figure 2.17 Performance Degradation achieved for watermarked MESA IP inserted with PD-HT (Sengupta et al., 2024) with variation in the attacker's selected number of inverters (X)

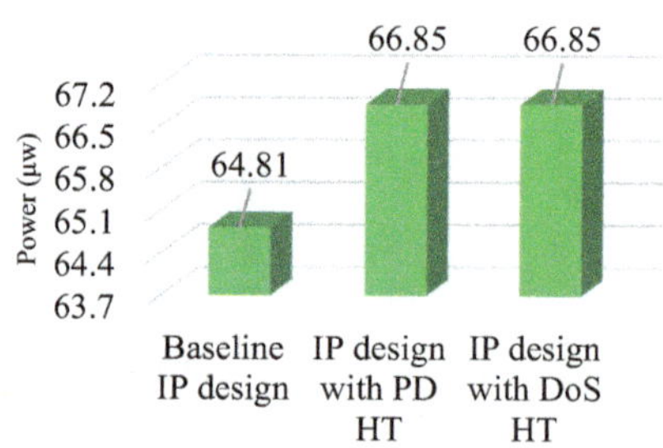

Figure 2.18 Power overhead due to insertion of PD-HT and DoS-HT (Sengupta et al., 2024) in watermarked MESA IP design

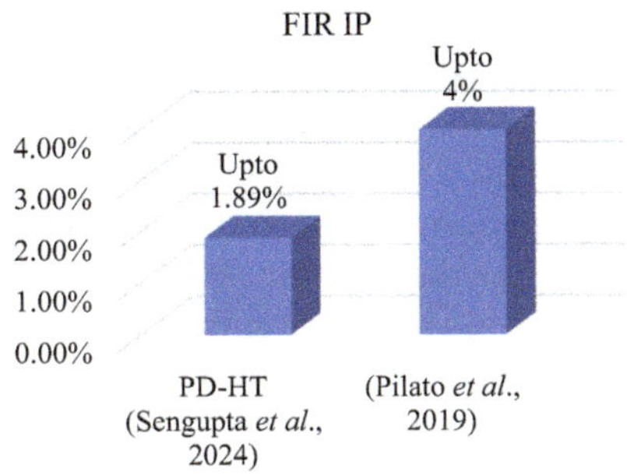

Figure 2.19 Area overhead comparison of the Trojan-infected IP design (with PD-HT) (Sengupta et al., 2024) with (Pilato et al., 2019)

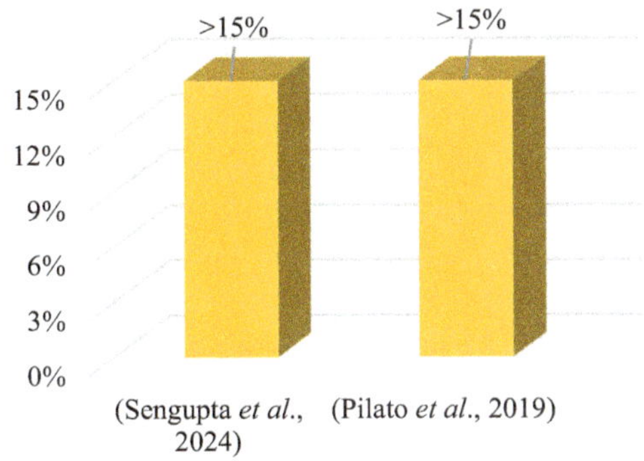

Figure 2.20 Comparison of (Sengupta et al., 2024) with (Pilato et al., 2019) in terms of performance degradation achieves

degradation also rises in a proportional manner. This behavior highlights the tunable nature of the attack, where an adversary can control the severity of the performance loss simply by adjusting the number of inverters secretly in the design. In practical terms, a smaller value of "*X*" allows the Trojan to remain stealthier with limited impact, whereas a larger value of "*X*" causes a more pronounced slowdown, thereby offering an attacker the flexibility of performance degradation achieved. For watermarked MESA IP, the performance degradation achieved is between 11.41% and 27.37% relative to a Trojan-free design, thereby demonstrating its strong disruptive capability despite the minimal area overhead. In addition to area and performance, Figure 2.18 highlights the power implications of the PD-HT and DoS-HT relative to Trojan-free IP design. Furthermore, a comparative perspective of the Trojan area overhead is provided in Figure 2.19, where the PD-HT inserted FIR IP design is evaluated against another state-of-the-art method (Pilato *et al.*, 2019). The results reveal that PD-HT achieves a smaller area overhead compared to (Pilato *et al.*, 2019). It is also worth noting that Pilato *et al.* (2019) have not performed Trojan insertion on HLS-generated watermarked IPs, while PD-HT and DoS-HT are explicitly designed to target such designs. Figure 2.20 further reinforces the capability of the PD-HT and DoS-HT by showing that it consistently achieves performance degradation greater than 15% for FIR IP. Finally, Figure 2.20 underscores the stealthiness of the PD-HT and DoS-HT. Even when subjected to recent detection techniques, the HLS-based PD-HT and DoS-HT successfully evade identification, highlighting the challenge they pose to existing countermeasures. Collectively, these results establish that the PD-HT and DoS-HT achieve a carefully balanced combination of low overhead, strong disruptive capability, and high stealthiness. The demonstration of how PD-HT can evade the recent detection techniques is explained in (Sengupta *et al.*, 2024).

2.9 Conclusion

In conclusion, this chapter has served as a comprehensive state-of-the-art reading for the HLS-aided hardware Trojan insertion technique and points of vulnerabilities within the HLS flow. Specifically, the malevolent HLS framework explained in this chapter exploited unutilized free ports in watermarked IPs, which an attacker can use to insert sneaky backdoor hardware Trojans. By leveraging these weak spots, an attacker can stealthily embed hardware Trojans, thereby transforming a regular IP design into a compromised one. Two representative classes of Trojans were discussed and analyzed: the PD-HT, which deliberately reduces system performance, and the DoS-HT, which affects operational execution. Beyond watermarked IPs, the methodology (Sengupta *et al.*, 2024) explained in this chapter is also extendable to non-watermarked IP designs, demonstrating its versatility and broad applicability. Through detailed analysis, this chapter has highlighted how even small-scale insertions can lead to significant consequences in terms of performance and security.

2.10 Questions and exercises

1. What are the different stages of HLS?
2. What is a PD-HT attack?
3. What is a DoS-HT attack?
4. How is the HLS Trojan triggered?
5. What is the utility of TSB in HLS Trojan Logic?
6. What is the use of the inverter chain in HLS Trojan logic?
7. How can an attacker enhance the impact of a PD-HT attack?
8. Name a few recent Trojan detection techniques.
9. What is the use of a comparator in an HLS Trojan?
10. How is GNN useful in hardware Trojan detection?
11. What is Black-HAT HLS?
12. What is the importance of overhead analysis during Trojan attacks?

References

A. Sengupta, A. Anshul, V. Chourasia, and N. Kumar, "M-HLS: malevolent high-level synthesis for watermarked hardware IPs," *IEEE Embedded Systems Letters*, vol. 16, no. 4, pp. 497–500, 2024.

S. Rizzo, F. Bertini, and D. Montesi, "Fine-grain watermarking for intellectual property protection," *EURASIP Journal on Information Security*, vol. 2019, p. 10, 2019.

S. Bhunia, M. Abramovici, D. Agrawal, *et al.*, "Protection against hardware Trojan attacks: towards a comprehensive solution," *IEEE Design Test*, vol. 30, no. 3, pp. 6–17, 2013.

A. Sengupta, S. Bhadauria, and S. P. Mohanty, "TL-HLS: methodology for low cost hardware Trojan security aware scheduling with optimal loop unrolling factor during high level synthesis," *IEEE Transactions on Computer-Aided Design of Integrated Circuits and Systems*, vol. 36, no. 4, pp. 655–668, 2017.

C. Pilato, S. Garg, K. Wu, R. Karri, and F. Regazzoni, "Securing hardware accelerators: a new challenge for high-level synthesis," *IEEE Embedded Systems Letters*, vol. 10, no. 3, pp. 77–80, 2018.

C. Pilato, K. Basu, F. Regazzoni, and R. Karri, "Black-Hat high-level synthesis: myth or reality?," *IEEE Transactions on Very Large Scale Integration (VLSI) Systems*, vol. 27, no. 4, pp. 913–926, 2019.

A. Sengupta, D. Roy, and S. P. Mohanty, "Triple-phase watermarking for reusable IP core protection during architecture synthesis," *IEEE Transactions on Computer-Aided Design of Integrated Circuits and Systems*, vol. 37, no. 4, pp. 742–755, 2018.

A. Sengupta, and M. Rathor, "Facial biometric for securing hardware accelerators," *IEEE Transactions on Very Large Scale Integration (VLSI) Systems*, vol. 29, no. 1, pp. 112–123, 2021.

A. Sengupta, R. Chaurasia, and A. Anshul, "Robust security of hardware accelerators using protein molecular biometric signature and facial biometric encryption key," *IEEE Transactions on Very Large Scale Integration (VLSI) Systems*, vol. 31, no. 6, pp. 826–839, 2023.

F. Koushanfar, I. Hong, and M. Potkonjak, "Behavioral synthesis techniques for intellectual property protection," *ACM Transactions on Design Automation of Electronic Systems*, vol 10, pp. 523–545, 2005.

B. L. Gal, and L. Bossuet, "Automatic low-cost IP watermarking technique based on output mark insertions," *Design Automation for Embedded Systems*, vol. 16, pp. 71–92, 2012.

J. Chen, and B. C. Schafer, "Watermarking of behavioral IPs: a practical approach," *2021 Design, Automation & Test in Europe Conference & Exhibition (DATE)*, Grenoble, France, pp. 1266–1271, 2021.

University of California Santa Barbara Express Group, accessed on Sept. 2023. [Online]. Available: https://web.ece.ucsb.edu/EXPRESS/benchmark/.

S. P. Mohanty, N. Ranganathan, E. Kougianos, and P. Patra, "High-level synthesis fundamentals," In: *Low-Power High-Level Synthesis for Nanoscale CMOS Circuits*, Springer, London, UK, 2008.

A. Sengupta, and S. P. Mohanty, *IP Core Protection and Hardware-Assisted Security for Consumer Electronics*, The Institute of Engineering and Technology (IET), pp. 1–20, 2019.

A. Sengupta, and D. Roy, "Antipiracy-aware IP chipset design for CE devices: a robust watermarking approach," *IEEE Consumer Electronics Magazine*, vol. 6, no. 2, pp. 118–124, 2017.

A. Sengupta, A. Anshul, V. Chourasia, and N. Bhui, "Security vulnerability (Backdoor Trojan) during machine learning accelerator design phases," *IT Professional*, vol. 27, no. 1, pp. 65–72, 2025.

A. Anshul, and A. Sengupta, "A survey of high level synthesis based hardware security approaches for reusable IP cores [Feature]," *IEEE Circuits and Systems Magazine*, vol. 23, no. 4, pp. 44–62, Fourthquarter 2023.

M. Abderehman, R. Gupta, R. R. Theegala, and C. Karfa, "BLAST: belling the Black-Hat high-level synthesis tool," *IEEE Transactions on Computer-Aided Design of Integrated Circuits and Systems*, vol. 41, no. 11, pp. 3661–3672, 2022.

M. Bushnell, and V. Agrawal, *Essentials of Electronic Testing for Digital, Memory and Mixed-Signal VLSI Circuits, Springer Publishing Company*, Incorporated, 2013.

A. Mondal, S. Karmakar, M. H. Mahalat, S. Roy, B. Sen, and A. Chattopadhyay, "Hardware Trojan detection using transition probability with minimal test vectors," *ACM Transactions on Embedded Computing Systems*, vol. 22, no. 1, Article 11, pp. 1–21, 2023.

M. Tehranipoor, and F. Koushanfar, "A survey of hardware Trojan taxonomy and detection," *IEEE Design Test of Computers*, vol. 27, no. 1, pp. 10–25, 2010.

X. Wang, M. Tehranipoor, and J. Plusquellic, "Detecting malicious inclusions in secure hardware: challenges and solutions," *IEEE International Workshop on HOST*, Anaheim, CA, USA pp. 15–19, June 2008.

M. Ludwig, A.-C. Bette, and B. Lippmann, "ViTaL: verifying Trojan-free physical layouts through hardware reverse engineering," *IEEE Physical Assurance and Inspection of Electronics (PAINE)*, Washington, DC, USA, 2021, pp. 1–8.

A. Mondal, M. H. Mahalat, S. Mandal, S. Roy, and B. Sen, "A novel test vector generation method for hardware Trojan detection," *32nd IEEE International System-on-Chip Conference (SOCC)*, Singapore, 2019, pp. 80–85.

K. Wu, and R. Karri, "Fault secure datapath synthesis using hybrid time and hardware redundancy," *IEEE Transactions on Computer-Aided Design of Integrated Circuits and Systems*, vol. 23, no. 10, pp. 1476–1485, 2004.

CAD for Assurance, Crypto-Steganography Tool, *IEEE CEDA and IEEE Hardware Security and Trust Technical Committee (HSTTC)*, accessed on Sept. 2023. [Online]. Available: https://cadforassurance.org/tools/ip-ic-protection/crypto-steganography-tool/.

CAD for Assurance, KHC-Stego Tool: Key-Triggered Hash-Chaining Driven Steganography Tool, *IEEE CEDA and IEEE Hardware Security and Trust Technical Committee (HSTTC)*, accessed on Sept. 2023. [Online]. Available: https://cadforassurance.org/tools/ip-ic-protection/khc-stego/.

CAD for Assurance, Faciometric Hardware Security Tool, *IEEE CEDA and IEEE Hardware Security and Trust Technical Committee (HSTTC)*, accessed on Sept. 2023. [Online]. Available: https://cadforassurance.org/tools/ip-ic-protection/faciometric-hardware-security-tool/.

F. Nejadmoghadam, A. Mahani, and Y. S. Kavian, "A new testing method for hardware Trojan detection," *Elsevier, Procedia Technology*, vol. 17, pp. 713–719, 2014.

K. C. Mohapatra, M. Priyatharishini, and M. Nirmala Devi, "Improving transition probability for detecting hardware Trojan using weighted random patterns," *Ubiquitous Communications and Network Computing. LNCS*, Springer, Cham, vol. 276, 2019.

A. Sengupta, and S. Bhadauria, "Untrusted third-party digital IP cores: power-delay trade-off driven exploration of hardware Trojan secured datapath during high level synthesis," *25th IEEE/ACM Great Lake Symposium on VLSI (GLSVLSI)*, Pennsylvania, New York, NY, USA, May 2015, pp. 167–172.

J. Rajendran, Z. Huan, O. Sinanoglu, and R. Karri, "High-level synthesis for security and trust," *Proceedings of the 19th IEEE International On-Line Testing Symposium*, vol. 2013, pp. 232–233, 2013.

C. Wolinski, K. Kuchcinski, and E. Raffin, "Automatic design of application-specific reconfigurable processor extensions with UPaK synthesis kernel," *ACM Transactions on Design Automation of Electronic Systems*, vol. 15, no. 1, Article 1, 36 pp., 2010.

Y. W. Huang, S. Bhunia, and P. Mishra, "Scalable test generation for Trojan detection using side channel analysis," *IEEE Transactions on Information Forensics and Security*, vol. 13, pp. 2746–2760, 2018.

M. Rathor, and A. Sengupta, "Revisiting Black-Hat HLS: a lightweight countermeasure to HLS-aided Trojan attack," *IEEE Embedded Systems Letters*, vol. 16, no. 2, pp. 170–173, 2024.

R. Yasaei, L. Chen, S.-Y. Yu, and M. A. A. Faruque, "Hardware Trojan detection using graph neural networks," *IEEE Transactions on Computer-Aided Design of Integrated Circuits and Systems*, vol. 44, no. 1, pp. 25–38, 2022.

Open Cell NanGate Library, 15 nm Open Cell Library. Available: https://si2.org/open-cell-library/, last accessed on Feb. 2024.

A. Sengupta, S. Bhadauria, and S. P. Mohanty, "Embedding low cost optimal watermark during high level synthesis for reusable IP core protection," *2016 IEEE International Symposium on Circuits and Systems (ISCAS)*, Montreal, QC, Canada, 2016, pp. 974–977.

A. Sengupta, and S. Bhadauria, "Exploring low cost optimal watermark for reusable IP cores during high level synthesis," *IEEE Access*, vol. 4, pp. 2198–2215, 2016.

Chapter 3

Time-bomb HLS-based hardware Trojan attack for performance degradation

Anirban Sengupta[1] *and Nabendu Bhui*[1]

High-level synthesis (HLS) is an important design paradigm used for application-specific integrated circuits (ASICs), reusable intellectual property (IP) designs, or custom hardware accelerators. However, research has shown that third-party HLS tools/HLS frameworks can be exploited for malicious intent due to their untrustworthy nature. Some specific case studies reported in the literature prove that back-door Trojan logic can be injected into the HLS design flow by an adversary in the HLS tool development house. Such malicious HLS frameworks have the capability to compromise an IP design [which is expected to be integrated in a system-on-chip (SoC) design or produced as a standalone ASIC]. This chapter describes a recently emerged HLS Trojan vector called "Time-Bomb Hardware Trojan (TB-HT)" that has the capability to produce a performance degradation payload. The chapter elaborates on its threat model, insertion technique, and time-sensitive triggering mechanism.

The rest of the chapter is structured as follows: Section 3.1 describes the introduction of the chapter; Section 3.2 discusses salient features of this chapter; Section 3.3 provides a detailed discussion on the state-of-the-art approaches to HLS Trojans; Section 3.4 discusses HLS Trojan threat model; Section 3.5 describes the process of time-bomb HLS Trojan insertion during HLS flow; Section 3.6 discusses the TB-HT attack scenario during HLS; Section 3.7 describes the analysis of case studies; and Section 3.8 concludes the chapter by summarizing its significance and inferences.

3.1 Introduction

Various modern electronic, multimedia, and healthcare devices/systems heavily depend on secure and customized computing processors/hardware that require computer-aided design (CAD) flow at different abstraction levels of intellectual property (IP) designs (Sengupta, 2017a, 2017b). This type of custom computing processor/hardware is referred to as hardware IP, which forms the core of the

[1]Deptartment of Computer Science and Engineering, Indian Institute of Technology Indore, India

computing system. A commonly used CAD process known as high-level synthesis (HLS), which is an industry standard for designing custom IPs (Anshul and Sengupta, 2023; Rostami *et al.*, 2014; Sengupta *et al.*, 2018). The HLS process includes various key steps in the design process/flow, such as scheduling, allocation, binding, interconnect design, finite state machine (FSM) design, and register transfer level (RTL) datapath design. Therefore, HLS-based IP designs are crucial in various electronic, multimedia, and healthcare devices. The insertion of Trojans by an adversary/attacker into HLS-generated hardware IPs can lead to significant security vulnerabilities and may critically affect performance degradation (PD) (Sengupta *et al.*, 2017; Pilato *et al.*, 2019; Abderehman *et al.*, 2022). A hardware Trojan (HT) generally refers to malicious logic that is secretly inserted into an IP design during the design or manufacturing stages, intended to remain concealed/silent until triggered/activated by specific rare events or conditions. Authors in Sengupta *et al.* (2017) and Pilato *et al.* (2019) have shown that an adversary/attacker can jeopardize an IP design by compromising an HLS tool with the aim of embedding/injecting a Trojan secretly in different stages, like HLS library, scheduling, allocation, interconnect design, and FSM design.

Depending on the kind of Trojan payloads secretly inserted into the HLS design process, it is classified into the following: performance degradation hardware Trojan (PD-HT) in the scheduling stage (Sengupta *et al.*, 2024), degradation hardware Trojan (DG-HT) in the scheduling stage (Pilato *et al.*, 2019), functional hardware Trojan (F-HT) in the HLS tool library (as components) (Sengupta *et al.*, 2017), battery exhaustion hardware Trojan (BE-HT) in the finite state machine (FSM) design stage (Pilato *et al.*, 2019), denial of service hardware Trojan (DOS-HT) in the interconnect design stage (Sengupta *et al.*, 2024), downgrade attack hardware Trojan (DA-HT) in HLS hidden directives (Pilato *et al.*, 2019), etc.

This chapter discusses a state-of-the-art HLS Trojan type, viz., time-bomb HLS Trojan (TB-HT), that is responsible for inducing significant PD payload (Sengupta and Bhui, 2025). This TB-HT has been shown to be covertly inserted in the mux-based interconnect design stage of an HLS process. It is found to be evasive by nature, as it can only be activated through a specific rare-event triggering condition (predetermined by an adversary/attacker).

3.2 Salient features of this chapter

The salient features of this chapter are as follows:

- This chapter discusses a state-of-the-art HLS Trojan type, viz., Time-Bomb (TB) triggered hardware Trojan that has the ability to induce a significant performance degradation payload (Sengupta and Bhui, 2025).
- This chapter elaborates on the TB-HT insertion technique during the multiplexer-based interconnect design phase of the HLS process.
- This TB-HT is found to achieve much higher performance degradation (PD) with lower area overhead compared to other state-of-the-art techniques (Pilato *et al.*, 2019).

3.3 Discussion on the state-of-the-art approaches to HLS Trojan

For designing and triggering an HLS-aided time-bomb hardware Trojan, a combination of register, counter, comparator logic, and mux switches was introduced in the state-of-the-art (Sengupta and Bhui, 2025). It discussed an alternative method for Trojan insertion to adversely affect the performance of a digital design. Moreover, no prior related approaches have been discussed on embedding/inserting a TB-HT during the HLS process, especially by exploiting the multiplexor (mux) interconnect design stage. Further, a functional HLS Trojan was covertly inserted in the HLS tool library by Sengupta *et al.* (2017). Also, they (Sengupta *et al.*, 2017) have presented a detection mechanism of F-HT by applying dual-modular redundancy (DMR)-based scheduling and a distinct IP vendor policy. In Pilato *et al.* (2019), the authors have explained BE-HT and DG-HT attacks, which can drain the power of an IP design and compromise the security of cryptographic IP cores. These Trojans are inserted during the scheduling phase of the HLS flow. Neither of these works (Sengupta *et al.*, 2017; Pilato *et al.*, 2019) discussed a PD-HT attack that is activated by a time-bomb trigger, nor have they addressed Trojan insertion through the mux interconnect design phase, as Sengupta and Bhui (2025) did. More specifically, authors in Pilato *et al.* (2019) have concentrated on Trojan insertion during scheduling or allocation but did not explore exploiting vacant mux ports during the mux interconnect stage. Pilato *et al.* (2019) addressed downgrade attacks by exploiting computation path alternatives at a specific point with empty FSM bubbles to induce latency overhead. Sengupta *et al.* (Sengupta and Bhui, 2025) have presented the PD-Trojan payload that exploits a mux interconnect design phase of HLS flow through active-high mode tri-state buffer (TSB) chains. This technique resulted in a significantly higher PD increase with minimal area overhead, in contrast to Pilato *et al.* (2019), which did not induce such a large increase in performance degradation. Abderehman *et al.* (2022) focused on Trojan detection rather than Trojan insertion by introducing a C-to-register transfer level (RTL) equivalence checking method to detect degradation attacks, DG attacks, and BE attacks, respectively. However, their research did not address any Trojan attack/insertion technique during the HLS process. Additionally, Rathor *et al.* (Rathor and Sengupta, 2024) have presented a high-level transformation (HLT)-based detection technique to identify BE-HT attacks (Pilato *et al.*, 2019). Generally, it offered a lightweight countermeasure for Black-HAT HLS to detect BE hardware Trojan attacks post-HLT with the help of an operation count. However, detection methods like Sengupta *et al.* (2017), Jin *et al.* (2008), Quartus Tool (2025), Abderehman *et al.* (2022), and Rathor and Sengupta (2024) were not found capable of identifying/detecting this time-bomb-triggered HLS hardware Trojan, as it did not impact the functional behavior of the IP design. Additionally, the nature of PD-Trojan is stealthy, as it activates only under rare, specific conditions and is triggered by a time-bomb mechanism. For instance, the detection approach (Jin *et al.*, 2008) differentiated between normal and Trojan-affected chips using path delay fingerprints. However, it did not address the presence of Trojan payloads. Moreover, authors (Jin *et al.*, 2008) were unable to detect the performance degradation focused on TB-HT, as it only affected the performance without influencing/affecting the functionality and/or data output. Moreover, as chip complexity increases, this method (Jin *et al.*, 2008) was found to be less effective, particularly in HLS-generated IP designs with complex data pipeline structures, large storage systems, and multilayered switching elements/units, where path delays and testing patterns grew exponentially. Furthermore, the detection approach in Sengupta *et al.* (2017) assumed that Trojans were inserted after netlist generation, whereas the approach explained in this chapter targeted Trojan insertion during the HLS phase, before RTL generation. In addition, Sengupta *et al.* (2023) described a detection/isolation technique based on triple-modular redundancy (TMR) datapath design for detecting functional Trojans, but this method was found ineffective for the time-bomb PD-Trojan, as it impacted performance without altering the functionality and/and data output. Moreover, Yasaei *et al.* (2025) presented a detection method assuming Trojans were inserted directly into the Verilog RTL/netlist by a rogue designer or third-party (3P)-EDA tool vendor. However, the TB-HT (Sengupta and Bhui, 2025) was injected during the pre-RTL synthesis stage of hardware IP design, making Yasaei *et al.*'s (2025) detection method ineffective. Additionally, the Pyverilog-based toolkit, used by Yasaei *et al.*, exhibited poor performance when learning to detect Trojans in complex HLS-based IP designs. Subsequently, the functional validation or simulation method used in (Quartus Tool, 2025) was also unable to detect this time-bomb-triggered PD-Trojan, as it did not affect the IP design's functionality after Trojan insertion. Therefore, the HLS-based TB-HT attack technique (Sengupta and Bhui, 2025) was secretive/stealthy, more powerful in performance degradation, and could not be detected using state-of-the-art detection techniques.

3.4 Discussion on HLS Trojan threat model

HLS Trojans take advantage of the automated features/characteristics of an HLS framework and present significant security risks. HLS creates RTL datapath straight from its high-level specifications [like C++/C/data flow graph (DFG)] and simultaneously reveals possible attack vectors. As mentioned in Sengupta *et al.* (2017), Pilato *et al.* (2019), and Abderehman *et al.* (2022), the HLS tool design process reveals significant security vulnerabilities, which are driven by two primary reasons. At first, this HLS design process may create hidden vulnerabilities, allowing the injection of malicious hardware Trojans during IP design, which ultimately jeopardizes the integrity of the developed IP or leads to inferior IP components being accessible in the market. Secondly, HLS tools/frameworks with hidden Trojans can be part of the national-level attack toolkit, which allows tools/frameworks developed in one country/nation to be used in another country/nation for designing SoCs. These HLS frameworks are highly untrustworthy and unreliable, and possibly contain compromised versions that may incorporate backdoor Trojans without the knowledge of the IP designers. A rogue HLS tool developer may gain access to the mux-based interconnect design phase of the HLS workflow, where she/he can inject/embed a backdoor Trojan during the development of the HLS tool. Therefore, an adversary/attacker situated within the HLS tool designer's/vendor's house can gain access to critical design phases of the HLS workflow, where she/he can exploit this privilege to secretly inject Trojan logic in the mux-based interconnection design phase. The similar kind of threat model and attacker profile has been elaborated in the state-of-the-art (Pilato *et al.*, 2019; Abderehman *et al.*, 2022; Sengupta *et al.*, 2024). These HLS tools can carry covertly inserted Trojans that can only be activated through a specific rare-event triggering condition (predefined by an adversary/attacker). Additionally, RTL IP designs generated by the HLS framework are significantly larger and much more complex than their higher abstraction level specifications, which creates challenges and difficulties for the designers/developers to understand them. Moreover, as HLS trojans are activated under specific pre-established rare events and are often stealthy and elusive in nature, this helps them to avoid detection mechanisms like side-channel analysis, equivalence checking, exhaustive testing, and functional validation. These HLS Trojans can introduce security vulnerabilities like performance degradation, data integrity issues, battery/power exhaustion, etc. in the developed IP design (Sengupta *et al.*, 2017; Pilato *et al.*, 2019; Abderehman *et al.*, 2022).

3.5 Process of TB-HT triggering and insertion during HLS flow

This chapter discusses a time-bomb-triggered HLS hardware Trojan that is capable of degrading the significant performance of a digital system, which an adversary/attacker can covertly insert by utilizing the vacant input ports in the mux-based interconnection design of the HLS flow (Sengupta and Bhui, 2025). Generally, in all the IP designs, details of the mux units are generated. In the IP datapath architecture, the generated muxes contain at least one vacant (free) port that can be easily exploited by an attacker/adversary to secretly insert Trojans. In this chapter, a time-bomb-triggered PD-HT indicates a malicious modification in the IP design that generates PD payload under a particular rare-event time-duration-based triggering circumstance/condition. This HLS Trojan utilizes a time-bomb trigger, meaning that the Trojan mechanism activates (by an attacker) only after a predefined time period has passed. This TB-HT trigger is designed to activate only when the modulus up-counter (Mod "N" up-counter) value reaches the specific pre-determined state value (pre-loaded value "K"), which is stored in the internal memory (or a register). Then, an attacker takes advantage of the mux-based interconnect design phase of the HLS process to secretly inject this TB-HT. Figure 3.1 (Sengupta and Bhui, 2025) illustrates the insertion stage of TB-HT inside the HLS framework. The major primary inputs of this Trojan-inserted HLS design process (shown in Figure 3.1) consist of high-level specifications (like C++/C/control data flow graph (CDFG) of the hardware application, the HLS module library, and the resource constraints or pragma directives. Here, the HLS library comprises a collection of preconfigured/predesigned blocks/components used to design the RTL datapath of the hardware IP core. As this TB-triggered HLS Trojan activates solely during a unique, predefined, rare event and solely impacts performance, detecting this Trojan is quite difficult. To explain and demonstrate this PD HLS Trojan attack scenario (in the next subsection), a convolution filter IP design (Figure 3.2) is utilized that is generated by an HLS process (Sengupta and Bhui, 2025; Sengupta and Chaurasia, 2022).

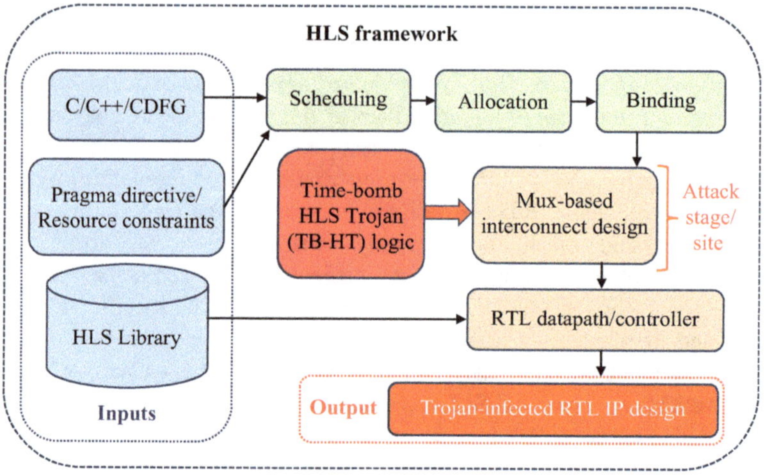

Figure 3.1 TB-HT insertion in the HLS design flow

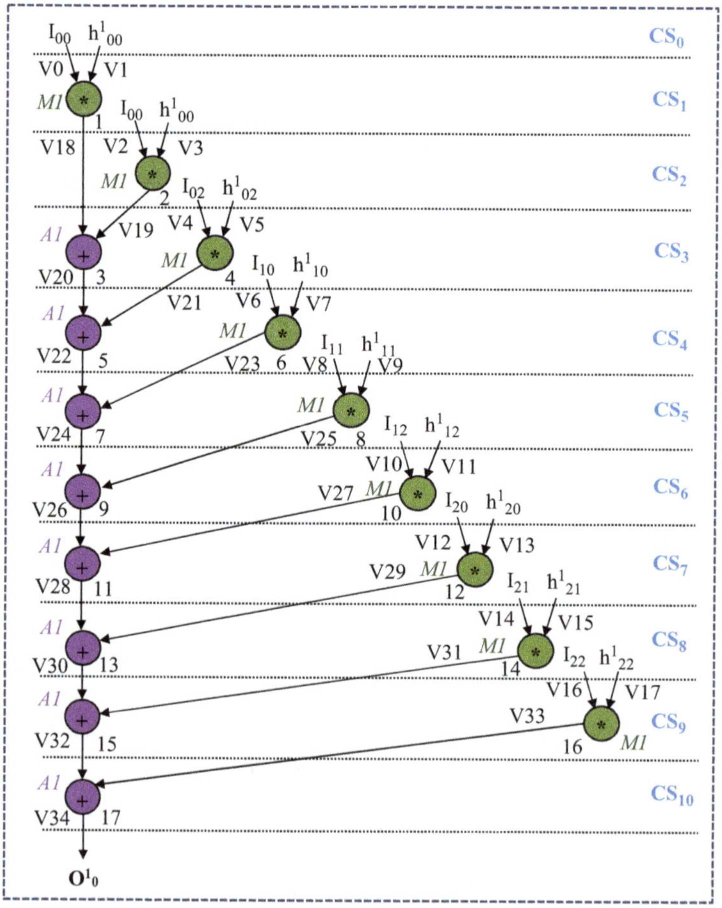

Figure 3.2 Scheduled data flow graph (DFG) of convolutional layer IP design (with a kernel of size 3 × 3) based on 1-multiplier, and 1-adder

3.6 TB-HT attack scenario during HLS

The TB-HT is inserted in the vacant input ports of the mux-based interconnection design phase of the HLS flow (Sengupta and Bhui, 2025). Figure 3.3 shows the top-level overview of the TB-HT-inserted IP design. A convolution filter IP design (Figure 3.2) is used to demonstrate the behavior of TB-HT, which is derived from Sengupta and Bhui (2025) and Sengupta and Chaurasia (2022) and generated by an HLS process. The SDFG (Figure 3.2) of convolutional layer IP design (with a kernel of size 3×3) is scheduled using 1-multiplier (M1), and 1-adder (A1) with the help of the list scheduling algorithm. The inputs of this SDFG are I_{00}, I_{01}, …, I_{22} and h_{00}^1, h_{01}^1, …, h_{22}^1; the output is O_0^1, and storage variables are V0, V1, V2, …, V34. Figure 3.4 demonstrates the time-bomb Trojan trigger logic that defines under what conditions the Trojan becomes activated and under what conditions the Trojan stays inactive. In this figure, red-colored logic indicates the TB triggering logic for performance degradation. The Trojan logic activates (by an attacker) only after a pre-loaded time period has passed. This TB-HT trigger is designed to activate only when the Mod "N" up-counter reaches the specific pre-defined state value "K," which is stored in the internal memory. The comparator ($N = K$) output signal is logic zero when Trojan is active; in other cases, it is logic one (when Trojan is inactive). If "$N = K$," then Trojan trigger logic sends an enable signal (i.e., @Sel.=0), which impacts the select line values of MUX ($S_0^M S_1^M S_2^M$) and DEMUX ($S_0^D S_1^D S_2^D$) and enables the tri-state buffer (TSB) chain.

Tables 3.1 and 3.2 represent the behavioral description of TB-HT with Trojan triggering possibilities (i.e., active state and inactive state), triggering conditions (i.e., N=K for active and $N \neq K$ for inactive/dormant), and their corresponding outputs. Here, the modulus "N" up-counter value is defined by the attacker/adversary, and pre-loaded values "K" are stored in the internal memory (or register). In case of a rare-event triggering, this time-bomb hardware Trojan is untraceable because it does not affect the functionality or nature of the IP design. This TB-HT is found to achieve much higher performance degradation with only a slight area overhead. In behavioral Table 3.1, it is mentioned that, for the convolutional filter IP datapath design, the select line values ($S_0^M S_1^M S_2^M$) of the 8:1 MUX are $S_0^M S_1^M S_2^M$=111, which means the Trojan is in an active state, while $S_0^M S_1^M S_2^M \neq 111$ means the Trojan is in an inactive state. Similarly, in Table 3.2, it is mentioned that, for the convolutional filter IP datapath design, the select line values ($S_0^D S_1^D S_2^D$) of the 1:8 DEMUX are as follows: $S_0^D S_1^D S_2^D = 111$ means the

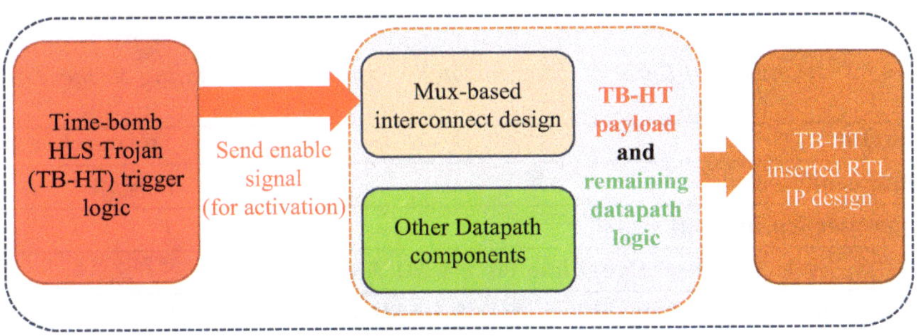

Figure 3.3 Top-level overview of the TB-HT-inserted IP design

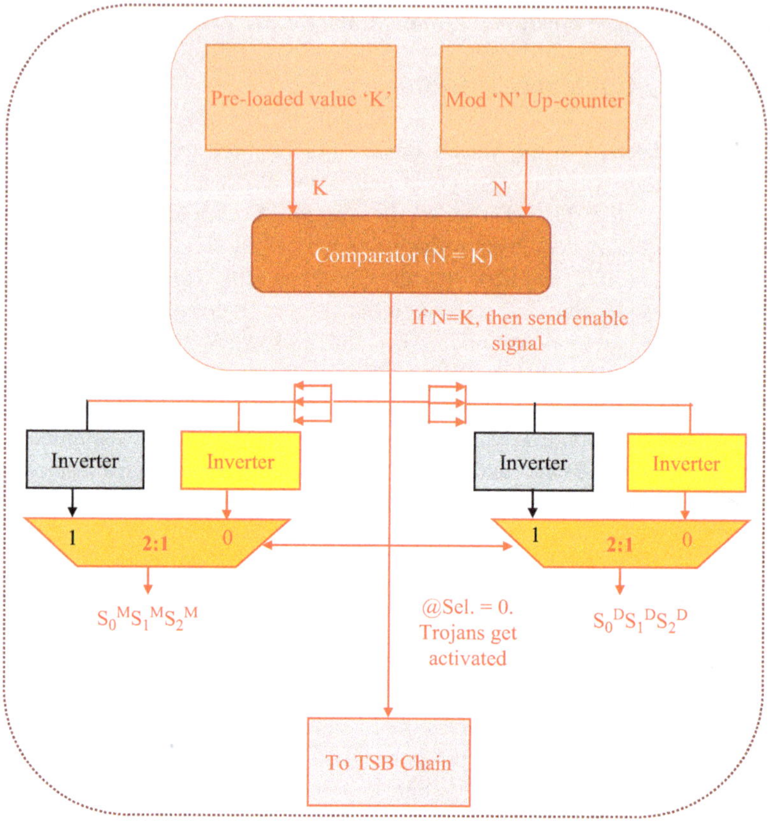

Figure 3.4 Time-bomb Trojan trigger logic

Table 3.1 Behavioral table of Trojan active state corresponding to the TB-HT

Possibilities	Trojan active
Trigger condition	N = K
Comparator o/p internal signal	0
$S_0^M S_1^M S_2^M$ **select internal signal line for 8:1 MUX**	$S_0^M S_1^M S_2^M = 111$
$S_0^D S_1^D S_2^D$ **select internal signal line for 1:8 DEMUX**	$S_0^D S_1^D S_2^D = 111$
O/P	Delayed of h_{00}^1

Table 3.2 Behavioral table of Trojan inactive state corresponding to the TB-HT

Possibilities	Trojan active
Trigger condition	$N \neq K$
Comparator o/p internal signal	1
$S_0^M S_1^M S_2^M$ **select internal signal line for 8:1 MUX**	$S_0^M S_1^M S_2^M \neq 111$
$S_0^D S_1^D S_2^D$ **select internal signal line for 1:8 DEMUX**	$S_0^D S_1^D S_2^D \neq 111$
O/P	Normal

Trojan is in an active state, while $S_0^D S_1^D S_2^D \neq 111$ means the Trojan is in an inactive state. The final output (in Table 3.1) indicates there is a delay in normal output (h_{00}^1), and the final output (in Table 3.2) indicates there is no delay in normal output, where h_{00}^1 defines the coefficient of the convolutional filter.

Here, Figure 3.5 demonstrates the TB-HT payload and remaining convolution filter IP datapath logic design. In this figure, red-colored logic/components indicate

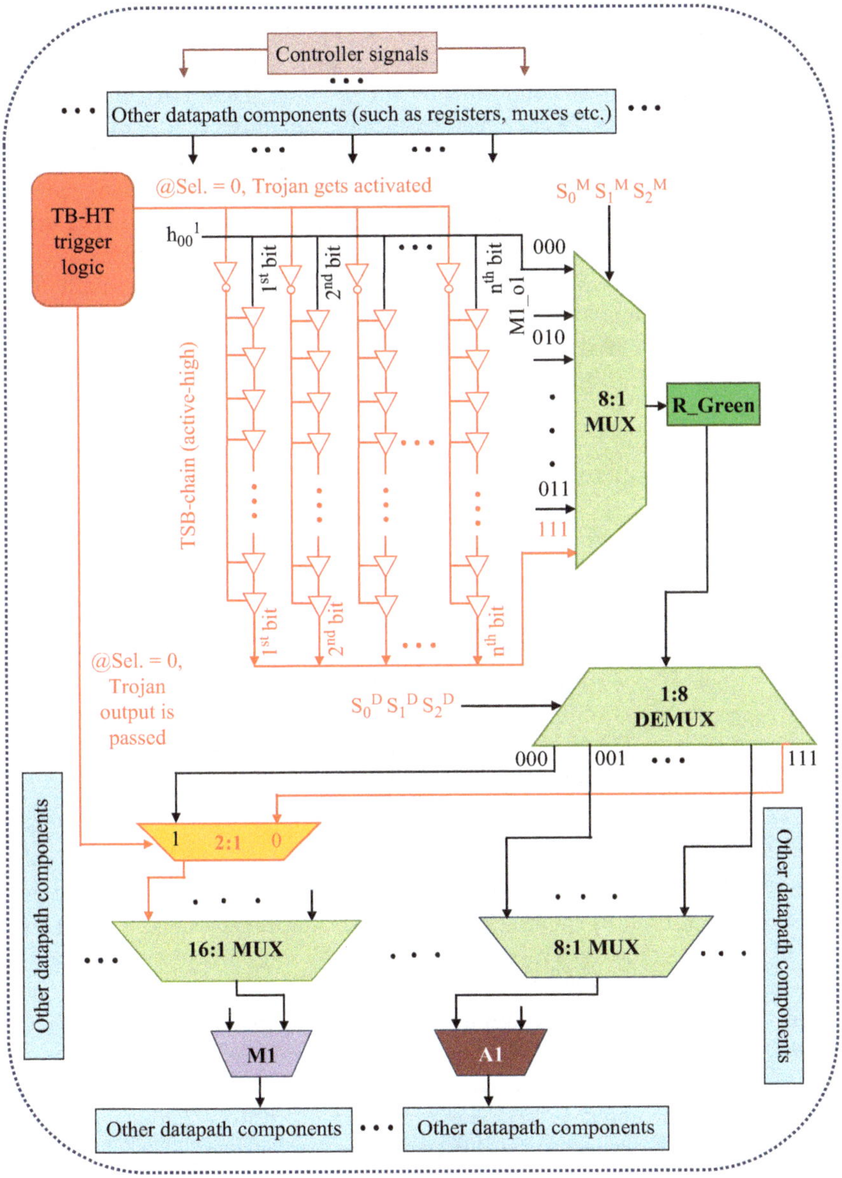

Figure 3.5 TB-HT payload and remaining convolution filter IP datapath logic design

the TB-triggered performance degradation hardware Trojan payload, and other non-red logics/components are the regular datapath designs of the IP. As indicated in Figure 3.5, on a particular rare-event triggering, the value "000" is automatically fed into all the inputs of 2:1 muxes (red-colored) to ensure $S_0^M S_1^M S_2^M = S_0^D S_1^D S_2^D = 111$ and to generate Trojan performance degradation payload (o/p). In other conditions (i.e., in a Trojan-free scenario), the IP datapath design works normally due to no Trojan trigger. As shown in Figure 3.5, TB-HT is inserted covertly into the free input ports of MUX (8:1), which is used in the convolutional IP datapath design. When the triggering logic gets activated (Trojan active state), i.e., @Sel.=0 and select line $S_0^M S_1^M S_2^M = 111$, then it passes through the active-high TSB chain, which is controlled by an inverter. This TSB chain insertion technique is found to incur higher performance degradation because of adding extra delay in the IP design. The length (L) of the secretly injected TSB chain is pre-defined by an adversary/attacker. The overall strength of the performance degradation payload achieved depends on this "L" value. If the TSB chain is inserted by an attacker and is larger in terms of "L" value, then the achieved PD payload is found to be greater. *Note:* This convolutional filter IP datapath design figure (Figure 3.5) focuses especially on highlighting the TB-HT payload, which is a part of the complete IP datapath design (Sengupta and Bhui, 2025; Sengupta and Chaurasia, 2022).

3.7 Analysis of case studies

In this chapter, the time-bomb-triggered HLS hardware Trojan technique is analyzed on some IP designs based on the following (Sengupta and Bhui, 2025):

- Performance degradation (PD) achieved due to the insertion of TB-HT in the target IP (from an attacker perspective—based on the TSB chain length).
- Area overhead due to insertion of TB-HT in the target IP (from an attacker perspective—based on the TSB chain length).
- Power overhead due to insertion of TB-HT in the target IP (from an attacker perspective—based on the TSB chain length).
- Comparative analysis of PD and area overhead between the TB-HT (Sengupta and Bhui, 2025) and Black-HAT HLS (Pilato *et al.*, 2019).

The TB-HT can achieve considerable PD, which may also change based on the TSB chain length used by an attacker/adversary within the datapath IP design. As shown in Figure 3.6, the TB Trojan-infected convolutional filter IP design reaches a PD of approximately ~19% (from ~2450.98 ps to ~2911.87 ps); the Trojan-infected sharpening filter IP design achieves a PD of approximately ~58% (from ~794.91 ps to ~1255.80 ps); the Trojan-infected JPEG-DCT reaches approximately ~50% (from ~927.40 ps to ~1388.28 ps); the Trojan-infected cardiac pacemaker results in about ~18.31% (from ~2517.23 ps to ~2978.11 ps); and the Trojan-infected ECG detector GLRT comes to a maximum of ~27.83% (from ~1656.07 ps to ~2116.95 ps).

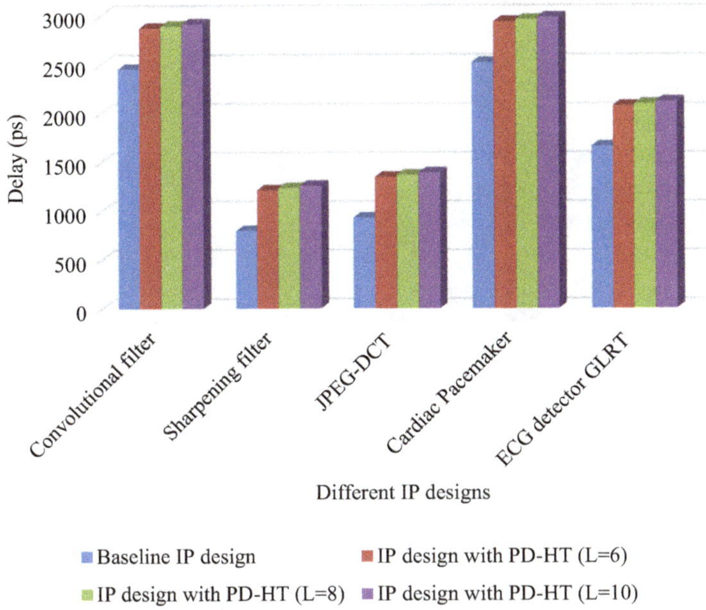

Figure 3.6 Performance degradation due to insertion of TB-HT in different IP designs with changes in attacker's selected TSB chain length (L)

Due to the achieved performance degradation, there are some areas of overhead in the respective target IP designs due to the TB Trojan insertion, which is shown in Figure 3.7. As shown in this figure, the TB Trojan-infected convolutional filter IP design incurs area overhead of approximately ~2.8% (from ~10448 gate count to ~10742 gate count); the Trojan-infected sharpening filter IP design incurs area overhead of approximately ~4.0% (from ~7264 gate count to ~7558 gate count); the Trojan-infected JPEG-DCT incurs area overhead of approximately ~3.3% (from ~8832 gate count to ~9126 gate count); the Trojan-infected cardiac pacemaker incurs area overhead of ~1.8% (from ~16096 gate count to ~16390 gate count); and the Trojan-infected ECG detector GLRT comes to a maximum of ~2.5% (from ~11744 gate count to ~12038 gate count). As shown in Figure 3.7, the area overhead resulting from the TB-HT insertion for target IP design is very low.

Similarly, while achieving PD, there is some power overhead in the respective target IP designs due to the TB Trojan insertion, which is shown in Figure 3.8. As shown in this figure, the TB Trojan-infected convolutional filter IP design incurs power overhead of approximately ~3.7% (from ~78.72 to ~81.67 μw); the Trojan-infected sharpening filter IP design incurs power overhead of approximately ~4.7% (from ~63.16 to ~66.10 μw); the Trojan-infected JPEG-DCT incurs power overhead of approximately ~4.48% (from ~65.70 to ~68.65 μw); the Trojan-infected cardiac pacemaker incurs power overhead of approximately ~2.26% (from ~130.38 to ~133.32 μw); and the Trojan-infected ECG detector GLRT

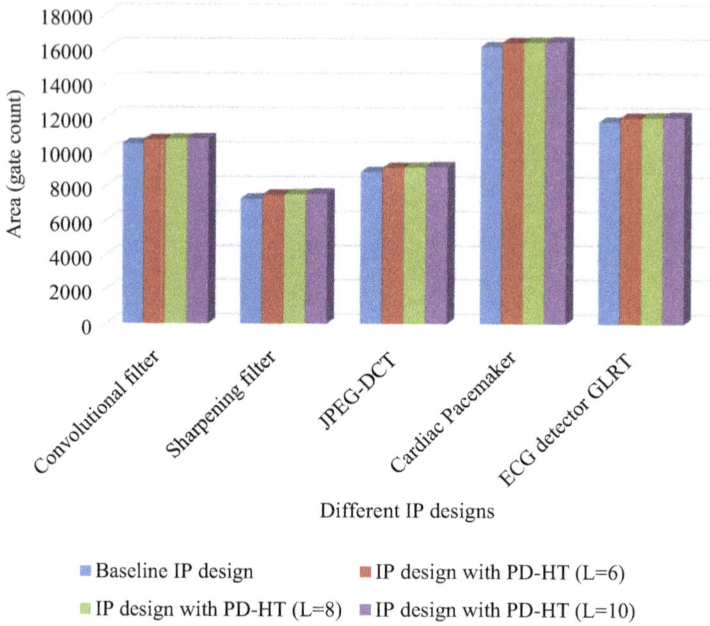

Figure 3.7 Area overhead due to insertion of TB-HT in different IP designs with changes in attacker's selected TSB chain length (L)

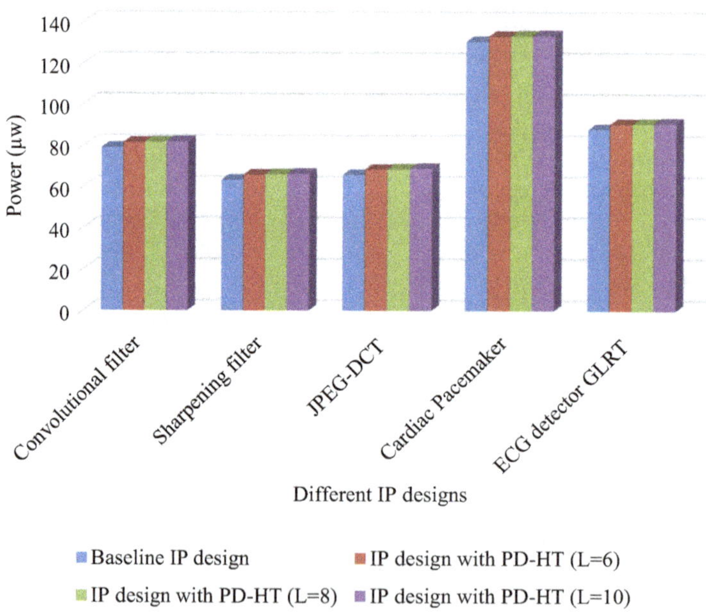

Figure 3.8 Power overhead due to insertion of TB-HT in different IP designs with changes in attacker's selected TSB chain length (L)

comes to a maximum of ~3.35% (from ~87.82 to ~90.76 μw). As shown in Figure 3.8, the power overhead resulting from the TB-HT insertion for target IP design is very minimal.

Further, as shown in Figure 3.9, the TB-HT-inserted IP designs (Sengupta and Bhui, 2025) were assessed in terms of performance degradation achieved in comparison to Black-HAT HLS (Pilato *et al.*, 2019). The assessment revealed that TB-HT-infected IP designs were capable of attaining higher performance degradation than Black-HAT HLS. This is because the TB-HT-inserted IP designs exploited the TSB inverter chain to incur performance degradation in comparison to Black-HAT HLS, which leveraged the insertion of bubble states/dummy states in the finite state machine design of the IP. For example, up to 58% performance degradation was attained by TB-HT insertion (Sengupta and Bhui, 2025) in comparison to only 17% by Black-HAT HLS (Pilato *et al.*, 2019). On a conservative note, a minimum of 19% PD was achieved by TB-HT insertion for convolutional filter IP design, which is still higher than the PD achieved by Black-HAT HLS.

Additionally, as shown in Figure 3.10, the TB-HT-inserted IP designs (Sengupta and Bhui, 2025) were assessed in terms of area overhead incurred in comparison to Black-HAT HLS (Pilato *et al.*, 2019). The analysis revealed that TB-HT-inserted IP designs were competent to attain significant performance degradation at lower area overhead than Black-HAT HLS. This is because the TSB chain exploited for performance degradation is lightweight in nature and consumes a smaller area. For example, a minimum of 1.8% area overhead was incurred with respect to cardiac pacemaker IP design, as compared to 4% for Black-HAT HLS.

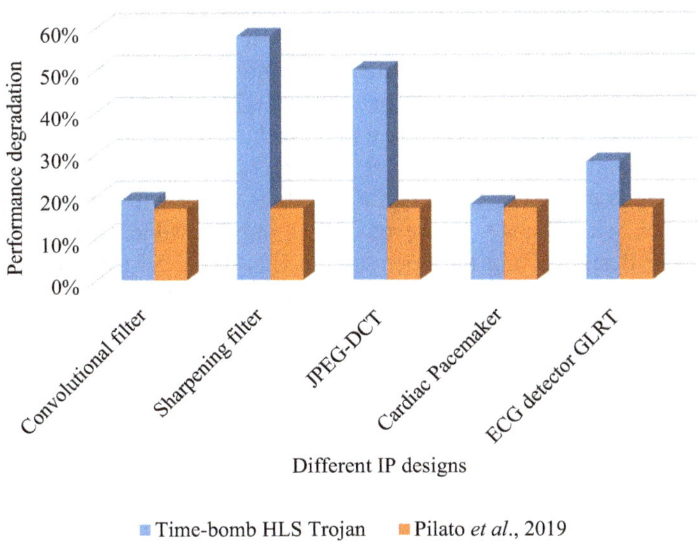

Figure 3.9 Performance degradation comparison between TB-HT and Pilato et al. (2019)

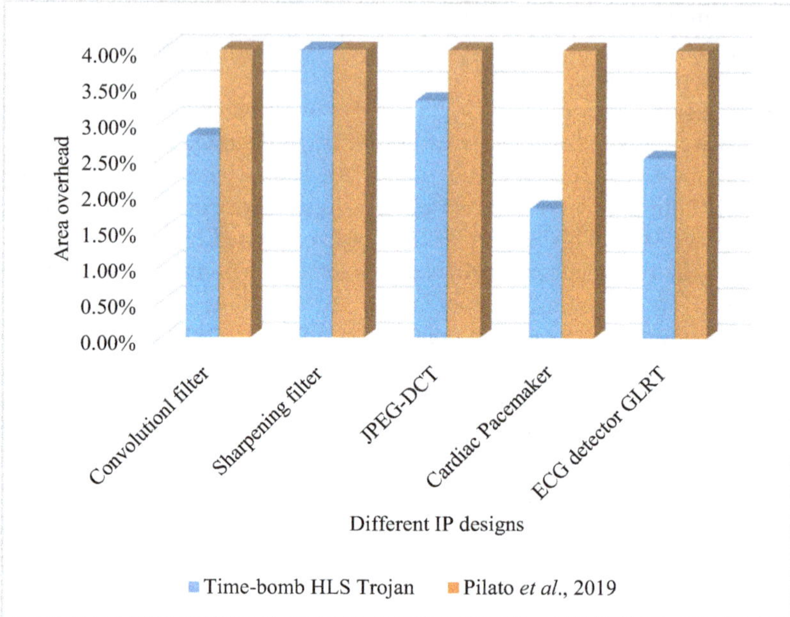

Figure 3.10 Area overhead comparison between TB-HT and Pilato et al. (2019)

Furthermore, this chapter also analyzes the evasion capability of TB-HT when evaluated against standard Trojan detection techniques (in Table 3.3). The TB-HT (Sengupta and Bhui, 2025) exhibits an elusive nature against the detection technique TL-HLS (Sengupta *et al.*, 2017). This is because the TL-HLS is based on the dual-modular redundancy (DMR) technique, which is only capable of handling HLS function Trojans. Therefore, it cannot handle TB-HT. Moreover, TB-HT also exhibits an evasive nature against a detection technique that is based on C to RTL equivalence checking (Abderehman *et al.*, 2022). This is because equivalence checking depends on the creation of a finite state machine with a data path (FSMD) for comparison with C-code. However, this is only possible when the TB-HT is inserted on IP designs using the open source Bambu HLS tool. On the contrary, TB-HT insertion was executed on a different publicly available HLS tool (Sengupta, 2025). Moreover, the detection technique using path delay fingerprint attempts to distinguish between Trojan-free and Trojan-infected IP designs (Jin *et al.*, 2008). However, it fails to successfully distinguish complex Trojan-infected HLS-generated IP designs. Additionally, TB-HT logic cannot be detected using the Trojan resistance technique (Sengupta *et al.*, 2023). This is because the Trojan resistance technique is based on the triple-modular redundancy (TMR) scheme, which is only capable of detecting and isolating HLS functional Trojans. Therefore, it is unable to detect TB-HT in IP designs. Moreover, a high-level transformation (HLT)-based detection technique (Rathor and Sengupta, 2024) is only able to detect battery exhaustion-based hardware Trojans (BE-HT). This is because BE-HT

Table 3.3 Demonstration of Trojan evasion for known detection techniques

Detection approach	**TL-HLS (Sengupta *et al.*, 2017)**	**C to RTL equivalence checking (Abderehman *et al.*, 2022)**	**HLT-based detection (Rathor and Sengupta, 2024)**	**Detection using path delay fingerprint (Jin *et al.*, 2008)**	**Trojan resistance (Sengupta *et al.*, 2023)**	**GNN (Yasaei *et al.*, 2025)**
Time-bomb Trojan status	Not detected	Not detected	Not detected	Not detected	Not detected	Not detected

(Note: HLT—High-level transformations from Rathor and Sengupta (2024), *DMR—dual-modular redundancy from* Sengupta *et al.* (2017), *TMR—triple-modular redundancy from* Sengupta *et al.* (2023)]

exploits the insertion of fake operations as Trojans during the HLS scheduling process. Therefore, this technique of Trojan insertion becomes detectable using HLT detection techniques. However, such techniques are not capable of detecting TB-HT logic, as it does not rely on the insertion of pseudo operations for the Trojan payload. Finally, a graph neural network (GNN)-based detection technique (Yasaei *et al.*, 2025) exploits the Pyverilog toolkit for Trojan detection in IP designs. The main limitations of such techniques are (a) they are incapable of handling VHSIC Hardware Description Language (VHDL) codes of Trojan-infected IP designs, and (b) they exhibit weaker learning behavior for complex HLS-generated IP designs.

3.8 Conclusion

This chapter provided an elaborate discussion on time-bomb HLS Trojan, viz., TB-HT, along with details of its threat model, insertion site, insertion technique, triggering mechanism, Trojan payload, and its evasion capability.

The chapter dives deep into the intricacies of TB-HT Trojan design during HLS design flow, where an attacker/adversary hiding in the HLS tool development house can covertly inject such Trojans in the design flow. The chapter also highlights the stealthy nature of such TB-HT Trojans by providing empirical details of performance degradation achieved and area/power overhead incurred. It also provides a comparison perspective with another similar HLS Trojan attack vector. Finally, the justifications of the elusive nature of TB-HT against standard detection techniques are also included. This chapter highlights the powerful nature of such time-sensitive TB-HT logic in the context of HLS IP designs.

3.9 Questions and exercises

1. How is TB-HT different from PD-HT?
2. How is TB-HT different from DoS-HT?

3. What are the different insertion sites for HLS Trojans?
4. How is Black-HAT HLS different from TB-HT?
5. How much is the maximum PD achieved for a cardiac pacemaker using TB-HT, and why?
6. Why is C to RTL checking not capable of detecting TB-HT?
7. How do you determine the time-sensitive nature of TB-HT?
8. How is the TSB chain different from the inverter chain in terms of activation?
9. What kind of Trojan can be detected using HLT techniques?
10. How is the area overhead estimated for TB-HT-inserted IP designs?

References

A. Sengupta, and N. Bhui, "Time-bomb HLS Trojan for performance degradation payload," *IEEE Design & Test*, vol. 42, no. 5, pp. 83–92, 2025, doi:10.1109/MDAT.2025.3575356.

A. Sengupta, "Hardware security of CE devices [Hardware Matters]," *IEEE Consumer Electronics Magazine*, vol. 6, no. 1, pp. 130–133, 2017a, doi:10.1109/MCE.2016.2614552.

A. Sengupta, "Hardware vulnerabilities and their effects on CE devices: design for security against Trojans [Hardware Matters]," *IEEE Consumer Electronics Magazine*, vol. 6, no. 3, pp. 126–133, 2017b, doi:10.1109/MCE.2017.2684940.

A. Anshul, and A. Sengupta, "A survey of high level synthesis based hardware security approaches for reusable IP cores [Feature]," *IEEE Circuits and Systems Magazine*, vol. 23, no. 4, pp. 44–62, Fourthquarter 2023, doi:10.1109/MCAS.2023.3325607.

A. Sengupta, S. Bhadauria, and S. P. Mohanty, "TL-HLS: methodology for low cost hardware Trojan security aware scheduling with optimal loop unrolling factor during high level synthesis," *IEEE Transactions on Computer-Aided Design of Integrated Circuits and Systems*, vol. 36, no. 4, pp. 655–668, 2017, doi:10.1109/TCAD.2016.2597232.

C. Pilato, K. Basu, F. Regazzoni, and R. Karri, "Black-Hat high-level synthesis: myth or reality?," *IEEE Transactions on Very Large Scale Integration (VLSI) Systems*, vol. 27, no. 4, pp. 913–926, 2019, doi:10.1109/TVLSI.2018.2884742.

M. Abderehman, R. Gupta, R. R. Theegala, and C. Karfa, "BLAST: belling the Black-Hat high-level synthesis tool," *IEEE Transactions on Computer-Aided Design of Integrated Circuits and Systems*, vol. 41, no. 11, pp. 3661–3672, 2022, doi:10.1109/TCAD.2022.3200513.

A. Sengupta, and R. Chaurasia, "Secured convolutional layer IP core in convolutional neural network using facial biometric," *IEEE Transactions on Consumer Electronics*, vol. 68, no. 3, pp. 291–306, 2022, doi:10.1109/TCE.2022.3190069.

M. Rathor, and A. Sengupta, "Revisiting Black-Hat HLS: a lightweight countermeasure to HLS-aided Trojan attack," *IEEE Embedded Systems Letters*, vol. 16, no. 2, pp. 170–173, 2024, doi:10.1109/LES.2023.3327793.

A. Sengupta, A. Anshul, V. Chourasia, and N. Kumar, "M-HLS: malevolent high-level synthesis for watermarked hardware IPs," *IEEE Embedded Systems Letters*, vol. 16, no. 4, pp. 497–500, 2024, doi:10.1109/LES.2024.3416422.

Y. Jin, and Y. Makris, "Hardware Trojan detection using path delay fingerprint," *2008 IEEE International Workshop on Hardware-Oriented Security and Trust*, Anaheim, CA, 2008, pp. 51–57, doi:10.1109/HST.2008.4559049.

Intel Quartus Tool. Accessed: October 2025. [Online]. Available: https://www.intel.com/content/www/us/ en/software-kit/666221/intel-quartus-ii-web-editiondesign- software-version-13-1-for-windows.html.

A. Sengupta, A. Anshul, and R. Chaurasia, "Exploration of optimal functional Trojan-resistant hardware intellectual property (IP) core designs during high level synthesis," *Microprocessors Microsystems*, vol. 103, 2023, Art. no. 104973, doi:10.1016/j.micpro.2023.104973.

R. Yasaei, L. Chen, S. -Y. Yu, and M. Abdullah Al Faruque, "Hardware Trojan detection using graph neural networks," *IEEE Transactions on Computer-Aided Design of Integrated Circuits and Systems*, vol. 44, no. 1, pp. 25–38, 2025, doi:10.1109/TCAD.2022.3178355.

A. Sengupta, D. Roy, S. P. Mohanty, and P. Corcoran, "Low-cost obfuscated JPEG CODEC IP core for secure CE hardware," *IEEE Transactions on Consumer Electronics*, vol. 64, no. 3, pp. 365–374, 2018, doi:10.1109/TCE.2018.2852265.

A. Sengupta, and R. Chaurasia, "Secure implantable cardiac pacemaker for medical consumer electronics," *npj Biomedical Innovations*, vol. 2, no. 1, p. 5, 2025, doi:10.1038/s44385-025-00008-y.

M. Rostami, F. Koushanfar, and R. Karri, "A primer on hardware security: models, methods, and metrics," *Proceedings IEEE*, vol. 102, no. 8, pp. 1283–1295, 2014, doi:10.1109/JPROC.2014.2335155.

A. Sengupta, *Crypto-Steganography Tool, IEEE CEDA*, https://cadforassurance.org/tools/ip-ic-protection/crypto-steganography-tool/, accessed on October 2025.

Chapter 4

Hardware Trojan attacks on HLS-based machine learning accelerators

Anirban Sengupta[1] *and Vishal Chourasia*[1]

In the current era, where the globalization of the design supply chain of integrated circuits plays a vital role, the security of intellectual property (IP) designs has become a crucial factor. The IP designs from third-party IP vendors are considered untrustworthy due to the possibility of backdoor hardware Trojans in them. This chapter focuses on describing hardware Trojan attacks on high-level synthesis (HLS)-based machine learning accelerator IP designs. In the present age, machine learning (ML) accelerators (hardware IPs) are significant due to their wide applicability. However, due to the involvement of third-party CAD tools such as HLS frameworks in the design process of such machine learning accelerators (hardware IPs), a potential backdoor Trojan attack (insertion) is highly likely. In this chapter, we show how Trojan attacks can be launched on ML hardware accelerators along with their Trojan payload.

The rest of the chapter is structured as follows: Section 4.1 provides the introduction of the chapter; Section 4.2 explains what an ML accelerator is and its potential security vulnerability. Section 4.3 provides a brief discussion on classes of Trojan attacks; Section 4.4 discusses the triggering mechanism of Trojan attacks on ML accelerators; Section 4.5 presents a discussion on Trojan detection; Section 4.6 discusses the analysis of case studies; and Section 4.7 concludes the chapter by summarizing its significance and implications.

4.1 Introduction

Machine learning (ML) is a field focusing on using the appropriate features to build models that can effectively perform specific tasks. At its core, the success of a machine learning algorithm depends on selecting the right features (data attributes) that help the model understand patterns and relationships within the data. Features can include a wide range of information, such as numerical values, categorical data, or even textual content. Additionally, machine learning can be

[1]Deptartment of Computer Science and Engineering, Indian Institute of Technology Indore, India

divided into three main types: supervised learning, unsupervised learning, and reinforcement learning.

(a) In supervised learning, the model is trained on a labeled dataset where each input has a relationship with the output. An example is email classification, where emails are pre-labeled as "spam" or "not spam," and the model learns to classify new emails based on this information.
(b) Unsupervised learning works with unlabeled data, aiming to uncover hidden patterns or groupings. For example, document clustering involves grouping similar documents based on their content, without any predefined labels.
(c) In reinforcement learning, an agent learns to make decisions by interacting with an environment and receiving feedback in the form of rewards or penalties. A common example is teaching an agent to play Go, where it learns from winning or losing and adjusts its strategy to maximize long-term rewards.

Moreover, machine learning algorithms are essential for solving a wide range of classification and prediction tasks, with the choice of algorithm depending on factors such as data type, task complexity, and available computational resources. Different high-computationally-intensive tasks, like image recognition, natural language processing, weather forecasting, or predictive analytics, require specific algorithms tailored to their needs.

Linear Regression (LR) is a simple and efficient algorithm for regression tasks with linear relationships. However, LR struggles with non-linear data and is sensitive to outliers. Also, LR is ideal for problems like forecasting temperature, data estimation, and predicting house prices.

Decision trees are easy to interpret and handle both categorical and numerical data. They work well for both classification and regression, but can overfit if not carefully tuned. They are great for tasks where model interpretability is important.

Support Vector Machines (SVMs) are powerful for high-dimensional and non-linear data, using kernel functions to create complex decision boundaries. They are computationally expensive and require careful hyperparameter tuning, but are effective for tasks like text classification and image recognition.

Random forests combine multiple decision trees to improve accuracy and reduce overfitting. They are robust and handle large datasets well, but are slower for predictions and less interpretable than single decision trees.

Convolutional Neural Networks (CNNs) are also a key component of machine learning, widely used for various application, such as image classification, segmentation, object detection, face recognition, analysis and prediction of energy usage and traffic load, voice analysis, and emotion detection, as shown in Figure 4.1 (Sengupta *et al.*, 2025; Jiang *et al.*, 2021; Gupta *et al.*, 2023; Sengupta and Chaurasia, 2022). Leveraging machine learning techniques, CNNs excel at processing visual and auditory data, enabling applications like security systems, autonomous vehicles, smartphones, and voice assistants. Their ability to learn complex patterns makes them crucial for improving user experiences, enhancing security, and enabling advanced functionalities in a variety of consumer technologies.

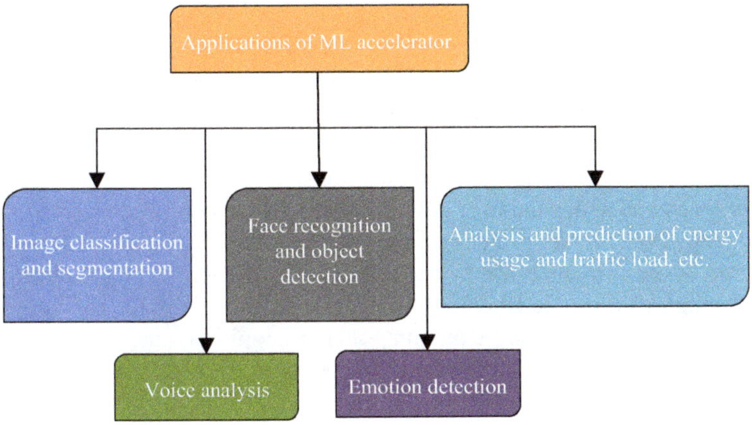

Figure 4.1 Different applications of ML accelerator

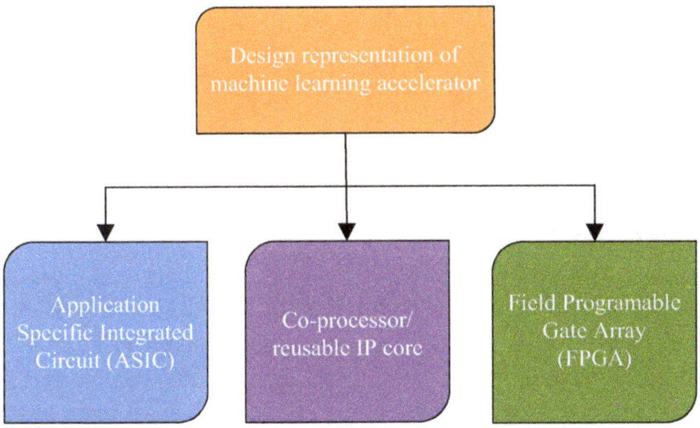

Figure 4.2 Alternative design representation for ML accelerators

The increasing computational demands of data-centric applications, such as machine learning, make it essential to utilize dedicated coprocessors/intellectual property (IP) designs/hardware accelerators to handle the intensive processing workloads. For instance, the logistic regression (LR) algorithm, which deals with large datasets during training, requires significant computational resources, especially when the data grows in size or complexity. To efficiently manage these resources, specialized computing platforms, including coprocessors/accelerators/hardware IP designs, field-programmable gate arrays (FPGAs), and application-specific integrated circuits (ASICs), are employed for alternative design implementation of machine learning algorithms, as shown in Figure 4.2 (Sengupta *et al.*, 2025; Gupta *et al.*, 2023; Sengupta *et al.*, 2023; Sengupta and Chaurasia, 2022). These dedicated accelerators are designed to optimize the power efficiency and performance for specific tasks, ensuring that the system meets the stringent requirements of high-speed data processing. These accelerators can be tailored to meet precise performance goals, such as minimizing energy consumption or maximizing processing speed, based on the specific needs of the application. For example, the convolutional layer IP core offers a flexible solution that can be integrated into various CNN-based applications. This IP core is particularly beneficial for portable and wearable devices, such as mobile phones, graphics processors, and Internet of Things (IoT) devices, where power efficiency and computational performance are key considerations (Sengupta and Rathore, 2020). By leveraging this reusable IP core, these devices can perform highly computation-intensive tasks, such as object detection, facial recognition, and augmented reality, with minimal energy consumption and enhanced processing speed, making them well-suited for deployment in a wide range of modern applications.

To formalize the development process of such ML accelerators, high-level synthesis (HLS) frameworks/tools are used. HLS frameworks accept high-level behavioral descriptions of ML algorithms and can automatically generate the corresponding register transfer level (RTL) datapath of the IP design (Sengupta *et al.*, 2025; Siemens Digital Industries Software, 2024; Sengupta *et al.*, 2023; Sengupta and Chaurasia, 2022). By using HLS, the ML accelerator development process becomes more efficient and adaptable, making it easier to meet the unique computational needs of various ML algorithms.

4.2 What is an ML accelerator and its potential security vulnerability?

ML hardware accelerators can play a crucial role in optimizing data-intensive tasks, such as image recognition, natural language processing (NLP), and autonomous driving. These accelerators are designed to significantly enhance the speed and efficiency of data processing, making them essential for real-time applications that require rapid analysis of large volumes of data. By offloading complex computations from general-purpose processors to specialized hardware, these accelerators enable faster execution of ML algorithms, leading to improved performance, reduced latency, and more efficient resource utilization.

Examples of these accelerators include the NVIDIA Deep Learning Accelerator (NVDLA), which is a scalable, open-source architecture optimized for deep learning workloads (Sengupta *et al.*, 2025; Gupta *et al.*, 2023). The linear regression (LR) hardware accelerator is another example, designed to speed up training tasks involving large datasets, often used in predictive analytics and classification problems (Sengupta *et al.*, 2023). The convolutional layer hardware accelerator is optimized for the convolutional operations in CNNs, critical for image and video processing tasks (Sengupta *et al.*, 2025; Sengupta and Chaurasia, 2022). These hardware accelerators are not only faster but also more energy-efficient compared to traditional CPU- or GPU-based processing, making them ideal for deployment in resource-constrained environments such as mobile devices, IoT applications, and embedded systems.

Machine learning accelerators, whether implemented as dedicated coprocessors, FPGAs, or ASICs, can be commonly developed using HLS flows that translate behavioral descriptions into RTL datapath. HLS is used for shortening development time and raising design productivity. The multiple automatic transformations and intermediate representations in the HLS/RTL toolchain create attack surfaces that malicious insiders or compromised CAD tools can exploit. At various stages of HLS, such as scheduling, allocation, interconnect (mux) generation, and RTL synthesis, an adversary can covertly inject backdoor hardware Trojans that remain dormant until triggered under rare events, making detection difficult and consequences severe.

4.2.1 Attack vectors and real-world impacts

Several real-world case studies demonstrate the potency of such attacks (Xue *et al.* 2020). For instance, Trojan logic can be used to bypass cryptographic units and leak secrets by enabling a rare-event (triggering) conditional bypass of the encryption logic, as illustrated in Figure 4.3. Similarly, power and side-channel leakage from ML accelerators can be leveraged to recover neural-network parameters. Figure 4.4 shows the side-channel security vulnerability of the NVDIA accelerator (NVDLA). Other documented attacks exploit device physics by inserting logic that induces negative-bias temperature instability (NBTI) stress on PMOS transistors (Kachave and Sengupta, 2019). An attacker can accelerate aging and degrade timing, speeding failure or permanently reducing the performance of DSP accelerator units.

Figure 4.5 shows the ML accelerator design process inside the HLS framework/design, comprising the scheduling, allocation, and mux-interconnect phases, which are particularly vulnerable. During scheduling, attackers can insert pseudo or fake operations (in underutilized functional units) that can consume extra energy/power during execution, manifesting as a battery-exhaustion attack. During the interconnect/mux design stage of HLS, unused mux inputs provide convenient hiding

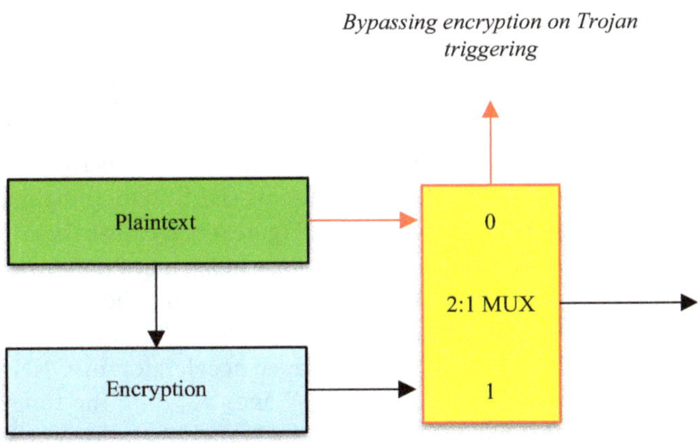

Figure 4.3 Trojan attack and security vulnerability in a cryptographic accelerator

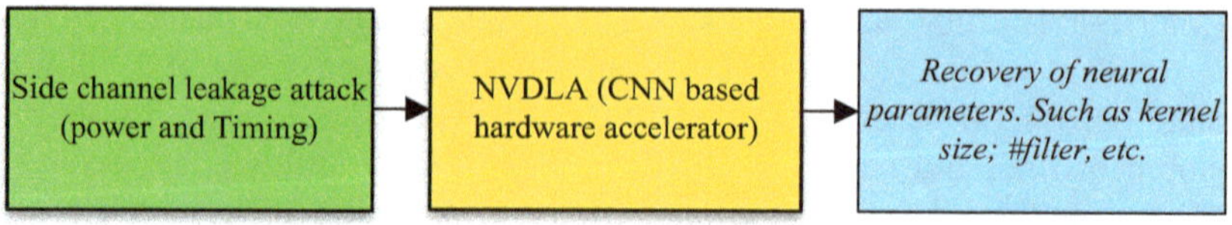

Figure 4.4 Example of a Trojan attack and the security vulnerability of a real-world application (NVDLA accelerator)

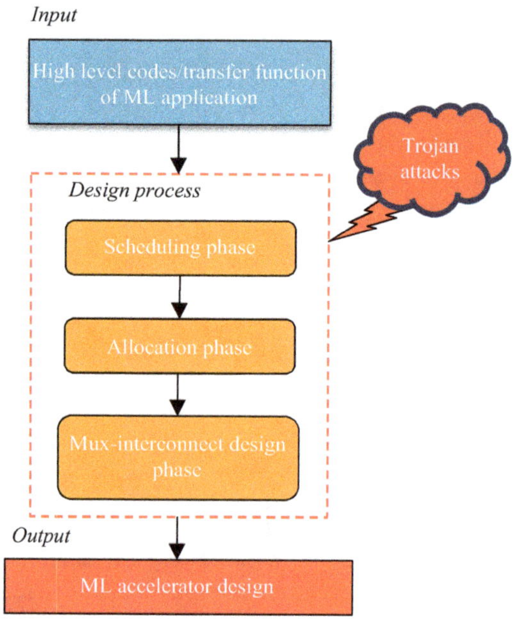

Figure 4.5 Potential Trojan attack during HLS-based ML accelerator design process

places for Trojan logic. An attacker who develops/controls the HLS tool can insert Trojan logic into such free ports, causing payloads that manifest later as denial-of-service Trojans (DoS-HTs), performance-degradation Trojans (PD-HTs), or data-damage Trojans (DD-HTs) (Sengupta *et al.*, 2023b). These hardware Trojans are inserted through compromised CAD software tools (HLS) during the machine learning accelerator design process, as illustrated in Figure 4.6. Because these Trojans often exploit unused or under-utilized resources without impacting the functionality of the ML accelerator design, they can evade standard functional validation.

Figure 4.7 illustrates the HLS design flow used to build an ML coprocessor/accelerator. For example, a convolutional layer accelerator in CNN (IP design) (Sengupta and Chaurasia, 2022). Here, "J_{MN}" and "K_{mn}" in the transfer function represent the input image of size M × N and the kernel of size m × n, respectively. Px denotes the output value of each element/pixel that corresponds to the output

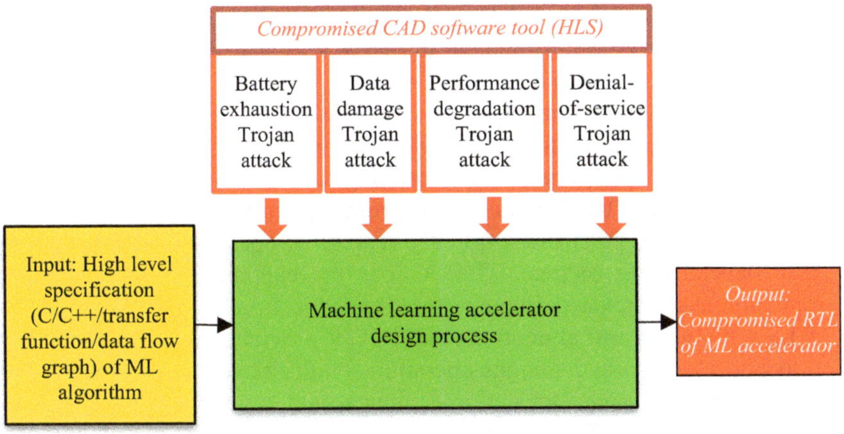

Figure 4.6 Established hardware Trojan attacks possible during ML accelerator designing

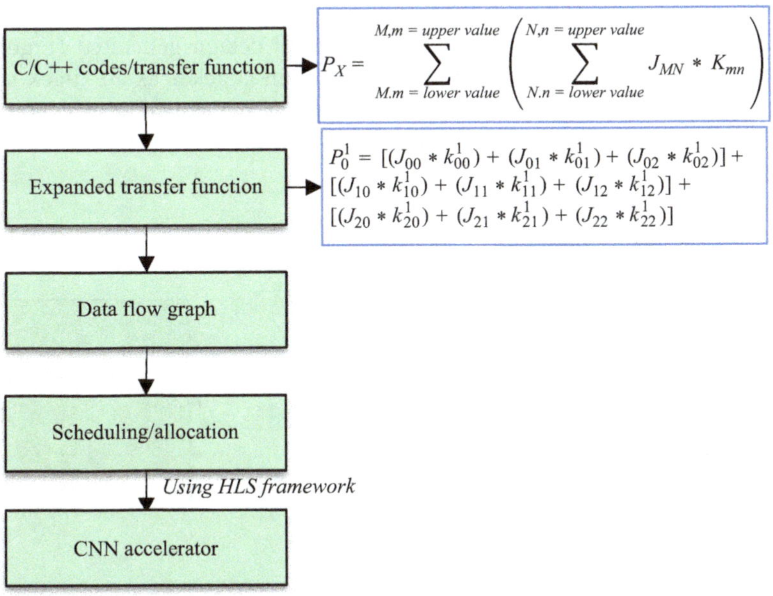

Figure 4.7 HLS design flow of ML (CNN) coprocessor/accelerator. mux: multiplexer

feature map; further in the expanded transfer function, each pixel value of the input image matrix and each kernel value of the kernel matrix "t" is represented by Jab and $\mathbf{k}_{pq}^{t}$ respectively. The flow proceeds as follows:

- **High-level specification (transfer function):** The process begins with the high-level description of the ML computation: the transfer function or algorithmic specification of the convolutional layer. This is the behavioral/functional representation input to the HLS tool, and the convolutional layer transfer function is shown in Figure 4.7.

- **Transfer-function expansion:** The high-level description is expanded or transformed into an explicit form that exposes the computational operations and data dependencies needed for mapping to hardware.
- **Data-flow graph (DFG) generation:** From the expanded transfer function, the tool constructs a DFG that explicitly represents operations (nodes) and their data dependencies (edges). The DFG is the canonical intermediate representation used for scheduling and resource allocation.
- **HLS scheduling and allocation:** The DFG is provided to the HLS scheduling/allocation stage, which assigns operations to clock cycles, maps operations to functional units, and allocates interconnect and storage resources (registers, multiplexers, etc.). This stage determines the microarchitectural timing and resource usage of the accelerator.
- **Datapath synthesis and RTL generation:** After scheduling and allocation, the HLS flow synthesizes the concrete RTL datapath and control logic that implements the scheduled operations.

The CNN convolutional layer datapath of IP design generated (Figure 4.8) shows unused ports, which can be exploited by an attacker as an attack surface. Therefore, if a CAD HLS tool used for designing CNN convolutional layer datapath IP is compromised, then an attacker can secretly exploit these free ports to

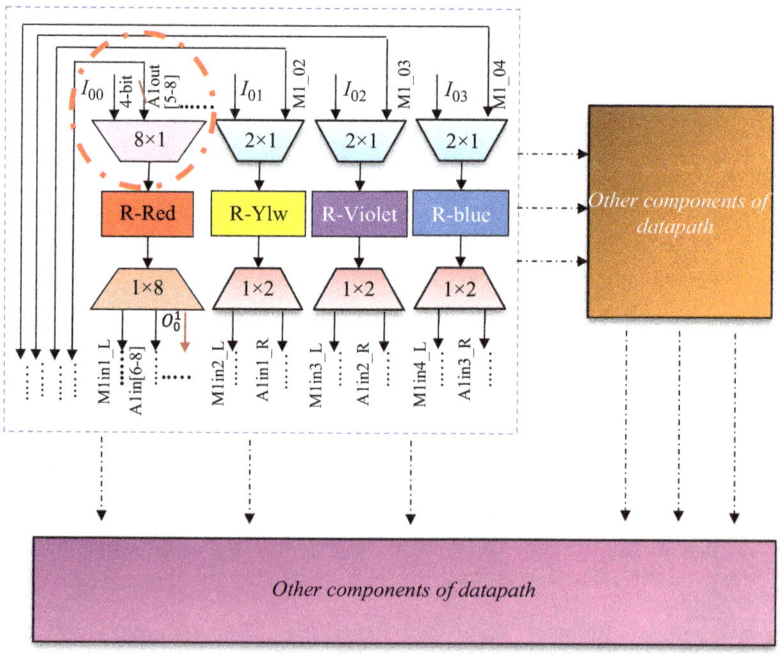

Figure 4.8 Partial view of CNN convolutional layer accelerator indicating kernel datapath for computing first output pixel O_o^1. Note: The complete view of the datapath is available at (Sengupta and Chaurasia, 2022).

insert hardware Trojans that would remain undetected (Sengupta *et al.*, 2025). Such malicious insertions can realize different payloads; for example, Trojans that cause denial-of-service (DoS HTs), performance-degradation (PD-HTs), or data-distortion (DD-HTs), each producing distinct adversarial effects on accelerator operation (Sengupta *et al.*, 2024).

A second exploitation vector is the HLS scheduling phase itself. An adversary can tamper with the scheduling process and can introduce pseudo-operations into the scheduled DFG that were not present in the original scheduled version. These fake operations can be mapped to unutilized hardware (functional resources such as adders/multipliers) for producing accelerated battery depletion (a battery-exhaustion attack) (Pilato *et al.*, 2019). Taken together, these vulnerabilities demonstrate that both the interconnect resources left unused in a generated datapath (e.g., free mux ports) and the HLS steps performed by the HLS tool (e.g., scheduling/allocation) can be maliciously exploited to hide backdoor Trojans and launch various attacks based on rare event triggering conditions. The specific mechanisms and impacts associated with each class of Trojan logic (DoS-HT, PD-HT, etc.) are described in detail in the following section.

4.3 Classes of Trojan attacks

In the context of ML accelerators, there are several types of Trojan attacks that have been widely recognized for their ability to compromise system performance and integrity. These attacks are classified as performance degradation Trojan (PD-HT), denial of service Trojan (DoS-HT), and battery exhaustion Trojan (BE-HT). These attacks have been carefully reviewed in the context of ML accelerator hardware, each exhibiting distinct characteristics that allow attackers to secretly infiltrate and manipulate system behavior. To better illustrate the mechanics of these Trojan attacks, we refer to the architecture of an ML accelerator, as depicted in Figure 4.9, which provides a generic framework for understanding how such attacks are launched into the system design. Figure 4.9 showcases an ML accelerator architecture under Trojan attack, providing a high-level view of the system structure. Along with Figures 4.9–4.11, provide more detailed insights into how specific Trojan variants, namely *PD-HT*, *DoS-HT*, and *BE-HT*, are introduced into the design of the ML accelerators (Sengupta *et.al.*, 2025).

This chapter focuses on various attack points, such as mux-based interconnects and unused/free ports, during the HLS-based ML accelerator design process. The mux-based interconnect design, which is central to the operation of many accelerators like the CNN convolution layer, is used in this explanation. Specifically, this design is shown in the context of ML accelerator IP design related to convolutional layers (Sengupta and Chaurasia, 2022). To highlight the insertion of Trojan logic, red-colored components are used in the illustrations of Figures 4.9–4.11, marking the places where the attacker has successfully inserted malicious logic. The red free port in an 8×1 mux represents an unused input port, which the attacker exploits for the covert insertion of the Trojan.

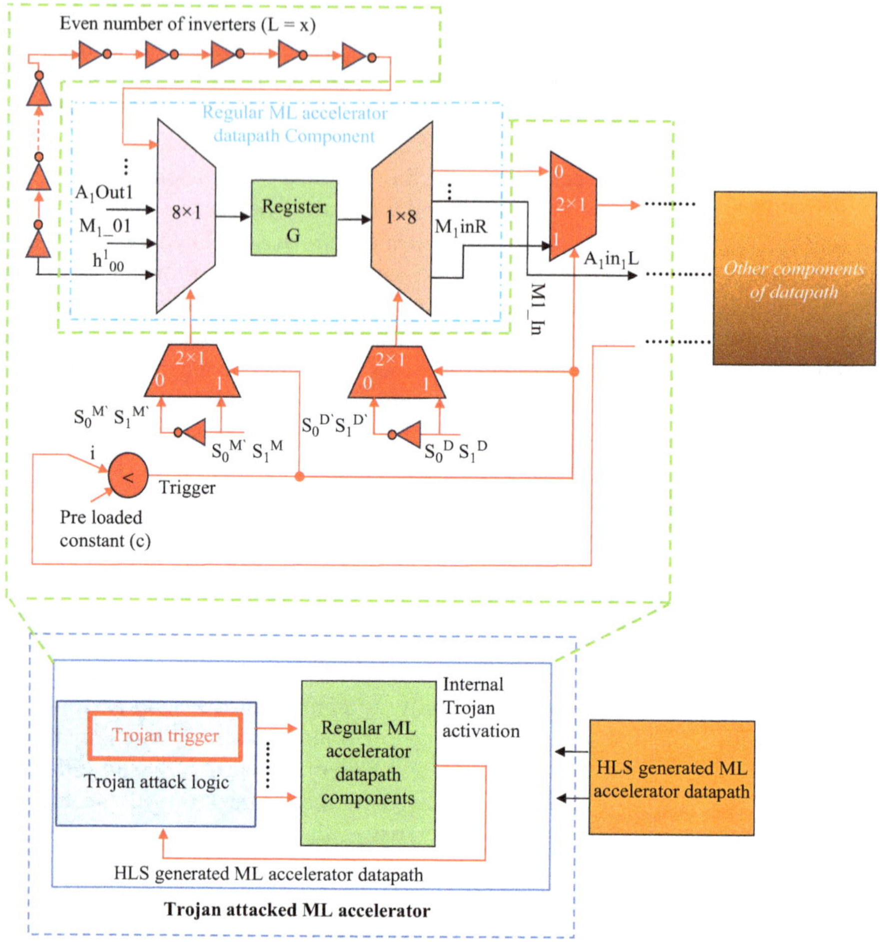

Figure 4.9 Performance degradation Trojan-inserted ML accelerator design

4.3.1 Performance degradation Trojan (PD-HT)

The PD-HT is one of the most subtle and insidious Trojan types, as it degrades the performance of the ML accelerator without altering its primary functionality. As illustrated in Figure 4.9, the *PD-HT* is inserted into the unused port of the mux interconnect design of the CNN convolutional accelerator IP. The triggering mechanism for this Trojan is tied to a rare and specific event/condition where a comparator inside the Trojan monitors the input signal "I," and when "i" equals a predefined (preloaded) constant "c," the Trojan is activated. Upon activation, the Trojan introduces unwanted delays into the system, slowing down the computational speed. Otherwise, under normal circumstances, when the trigger condition is not met, the system operates without any delay, making the Trojan very difficult to detect. The attacker can also control the degree of performance degradation by

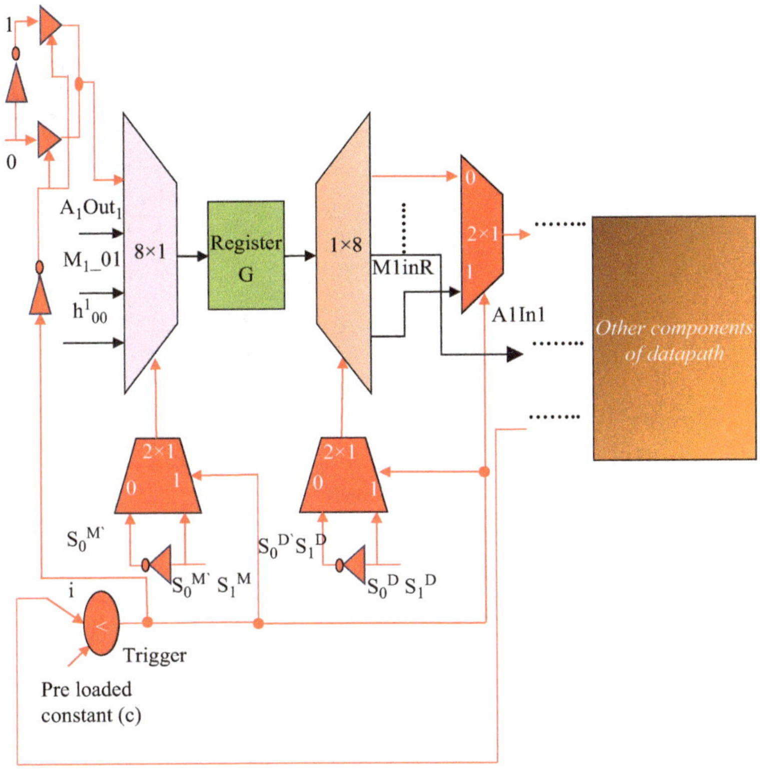

Figure 4.10 Denial of service (DoS) Trojan inserted into the ML accelerator design

manipulating the number of inverters used in the Trojan logic. Because these delays do not alter the expected outputs of the accelerator, traditional detection methods such as equivalence checking or functional validation are often ineffective at identifying this form of attack (Sengupta *et al.*, 2025).

4.3.2 Denial of service Trojan (DoS-HT)

Similar to the PD-HT, the DoS-HT Trojan is introduced via the unused port of the mux interconnect design of the convolutional layer IP, as shown in Figure 4.10. However, instead of degrading performance as in *PD-HT*, the *DoS-HT* seeks to disrupt or deny service during execution of the ML accelerator. When activated, it causes a sudden service denial of the ML accelerator. This type of Trojan can have serious consequences, especially if it affects critical hardware components within the system. Like the *PD-HT*, the *DoS-HT* is also triggered under a rare event/condition where the input "i" matches the constant "c." When the condition is met, the Trojan forces the accelerator into an unpredictable state, causing it to behave abnormally. Outside of this rare event, the accelerator behaves as expected, making detection very difficult (Sengupta *et al.*, 2025).

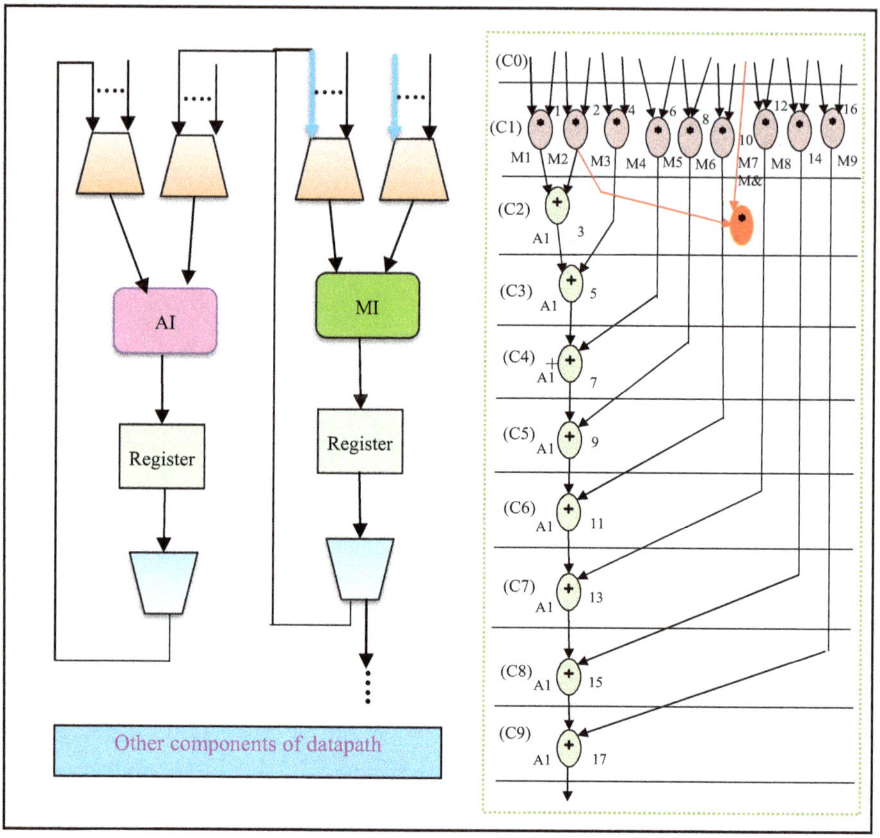

Figure 4.11 *Battery exhaustion Trojan attack in CNN convolutional layer accelerator*

4.3.3 Battery exhaustion Trojan (BE-HT)

Finally, the *BE-HT*, as shown in Figure 4.11, is particularly dangerous in embedded systems where power consumption is a critical concern. This Trojan is designed to drain the battery life of the system by causing excessive power consumption during operation. As illustrated in Figure 4.11, the unused multiplier functional unit (FU) resource is exploited by inserting a fake, Trojan-induced operation during the scheduling of DFG (control step two shown in red color). This insertion was carefully executed without breaching the predefined resource constraints, which include IP vendor resource constraints, e.g., nine multipliers and one adder. The methodology demonstrates how unutilized hardware resources within a design can be maliciously repurposed to insert stealthy Trojans. Moreover, similar Trojan operation insertions can also be carried out in other control steps by taking advantage of idle multiplier FUs, thereby increasing the potential for covert manipulation while maintaining the overall resource and timing constraints of the original design. Its impact is felt when the system begins to consume unwarranted amounts of power. This could result in a device running out of battery far sooner than expected, disrupting the user experience or causing the system to shut down prematurely. Like the other Trojans, BE-HT is activated under very specific rare conditions, making it stealthy and hard to detect during standard validation (Sengupta *et al.*, 2025; Pilato *et al.*, 2019).

These Trojan attacks are a stark reminder of the vulnerabilities that exist in hardware systems, particularly those used for ML accelerators. The sophisticated nature of these attacks, where performance degradation, denial of service, and battery exhaustion occur in subtle, infrequent ways, makes them particularly challenging to detect and mitigate. Each Trojan operates by exploiting unused or underutilized components in the system design, with attackers taking advantage of seemingly innocuous points in the data path. In this way, these Trojans are able to bypass traditional detection mechanisms, making them highly effective tools for those with malicious intent. By thoroughly understanding the underlying principles of these Trojan attacks, researchers and engineers can better design ML accelerators that are resilient to such threats, ensuring the continued security and reliability of next-generation hardware systems.

4.4 Triggering mechanisms of Trojan attacks on ML accelerator

In the context of malicious hardware threats, attackers employ a sophisticated method known as *stealthy HLS Trojan insertion*, where the Trojan remains dormant within a system until it encounters a very specific, rare condition. This activation condition, predetermined by the attacker, is secretly encoded into the system's design, making the Trojan virtually invisible during normal operations. This tactic lies in the use of comparator logic in the HLS datapath design stage, a mechanism that activates the Trojan's payload only when the system's state meets the exact rare condition set by an attacker. As a result, these malicious components evade detection, lying in a wait state to execute their harmful effects when the right trigger is finally encountered.

The process begins with the attacker programming a constant value, referred to as "c," into the IP design memory. This value is electrically programmable, meaning that it cannot be deleted easily without specialized techniques (Sengupta *et al.*, 2024). Figures 4.9 and 4.10 illustrate how this works: the first input to the comparator is linked to an internal functional unit (FU) of the system, such as an adder/multiplier. The second input is connected to the preloaded constant "c," which is preloaded in the memory. As the system operates, the comparator continuously compares the internal state (denoted as "i") against this stored preloaded value. Once the internal state exactly matches the value "c," the Trojan is triggered, unleashing its malicious payload into the system. The specific condition under which the Trojan activates varies based on the type of Trojan. For example, in the case of the *PD-HT* and *DoS-HT*, the trigger condition is the same when the internal state aligns with the pre-loaded value. However, the *BE-HT* Trojan has a distinct design, wherein it is activated after the counter value reaches a particular threshold, adding an additional layer of complexity to its activation. This subtle difference makes the *BE-HT* particularly tricky to identify, as it does not rely on the internal state but instead operates on an internal counter.

These hardware Trojans pose an alarming threat to the security and integrity of machine learning accelerator designs. Their ability to stay hidden until triggered by very specific conditions makes them exceedingly difficult to detect during routine operations or traditional system testing. Beyond their technical impact, these Trojans undermine the trust between ML accelerator IP vendors and the system integrator. The presence of such malicious elements can tarnish the reputation of the IP vendor and ultimately create vulnerabilities in ML infrastructures that could be exploited for malicious purposes. In the broader context, these threats underscore the importance of developing more sophisticated detection methods, as well as a more secure and transparent hardware supply chain, to ensure the reliability and trustworthiness of critical ML accelerator designs.

4.5 Discussion on Trojan detection

This section analyzes the state-of-the-art Trojan detection techniques. As we have understood so far, in the context of ML accelerator designs, HLS Trojans play a pivotal role. Therefore, it is essential to assess the utility of current Trojan detection techniques on Trojan-infected ML accelerator designs (Gubbi *et al.*, 2023). The Trojan-infected ML accelerator designs are often way more sophisticated and convoluted than their high-level algorithmic description. This makes it tedious to create an equivalence between the HLS-generated ML accelerator RTL and its original specifications counterpart. Therefore, detection techniques such as C to RTL (Abderehman *et al.*, 2022) equivalence checking do not work for PD-HT inserted ML accelerator designs and DoS-HT accelerator designs. However, it has the capability to detect BE-HT, as such Trojans are designed and implemented in the BAMBU HLS tool, where finite state machine datapath extraction is possible. Additionally, TL-HLS (Sengupta *et al.*, 2017), which is a DMR-based detection technique, is only capable of functional Trojan detection in HLS-generated IP designs. Moreover, detection techniques using side channel analysis (Huang *et al.*, 2018) are not effective against PD-HT and DoS-HT inserted designs. This is because these covert Trojan insertions are subtle and do not leak significant side channel information such as power, timing, etc. Further, a detection technique based on path delay fingerprint (Jin and Makris, 2008) attempts to distinguish between Trojan-free and Trojan-infected ML accelerator designs. However, such a technique becomes non-pragmatic when dealing with complex HLS-generated ML accelerator designs. Additionally, detection techniques that exploit graph neural networks (Yasaei *et al.*, 2022) are weak in detecting ML accelerator Trojans due to their inferior learning behavior for complex HLS-generated IP designs. Furthermore, a high-level transformation (HLT)-based detection technique (Rathor and Sengupta, 2024) is capable of detecting BE-HT attacks on ML accelerator designs and is not capable of detecting PD-HT and DoS-HT attacks. This is because the HLT-based detection technique exploits the comparison of the number of operations in the scheduled graph. Therefore, any pseudo-operation (indicating a fake Trojan operation) inserted in the scheduled graph is detected. A summary of these techniques is highlighted in Table 4.1.

Table 4.1 Analysis of different detection techniques on Trojan-infected ML accelerator designs (Note: "X" indicates "not detectable")

Different detection techniques	**C to RTL equivalence checking (Abderehman *et al.*, 2022)**	**TL-HLS (DMR-based security-aware scheduling) (Sengupta *et al.*, 2017)**	**Side channel analysis (Huang *et al.*, 2018)**	**Detection using path delay fingerprint (Jin and Makris, 2008)**	**GNN-based detection (Yasaei *et al.*, 2022)**	**HLT-based detection (Yasaei *et al.*, 2022)**
Performance degradation Trojan attack	X	X	X	X	X	X
Denial of service Trojan attack	X	X	X	X	X	X
Battery (power) exhaustion Trojan attack	✔	X	X	X	X	✔

4.6 Analysis of case studies

This section delves into an in-depth analysis of various ML accelerator architectures as discussed in recent case studies. Figures 4.12 and 4.13 provide a comprehensive overview of how the design area and latency evolve for the LR-ML accelerator when subjected to varying numbers of datasets (denoted as N). The key observation here is that both design area and latency demonstrate a direct correlation with the number of datasets processed by the accelerator. In other words, as the complexity of the input data grows, the hardware required to handle the computations expands proportionally, leading to increased design area and higher latency.

Figures 4.14 and 4.15 focus on a different ML architecture, i.e., the convolutional neural network (CNN) accelerator. Specifically, these figures highlight the

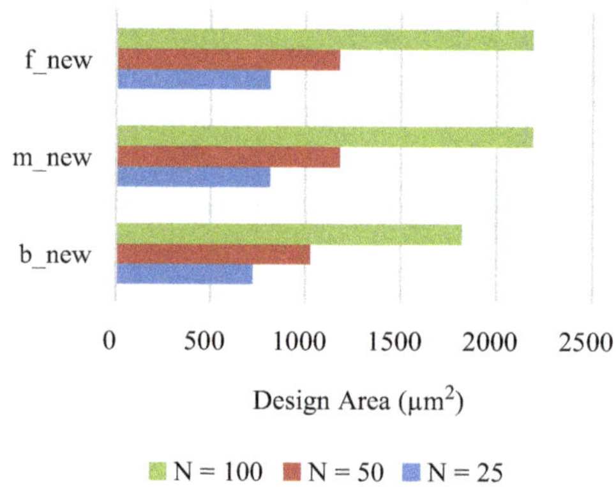

Figure 4.12 Design area for LR-ML accelerator corresponding to different numbers of datasets (N) (Sengupta et al., 2024, 2023)

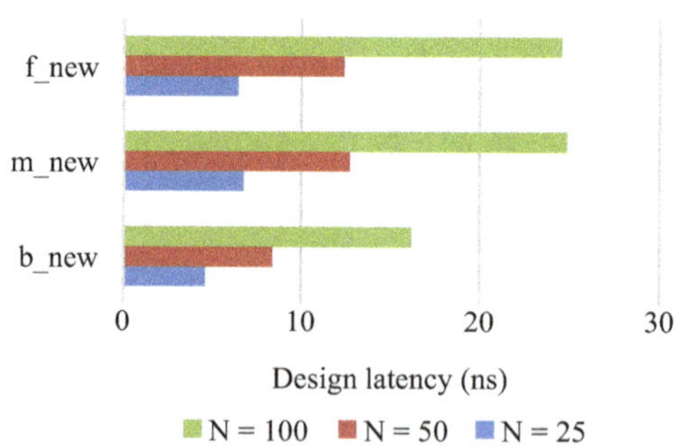

Figure 4.13 Design latency for LR-ML accelerator corresponding to different numbers of datasets (N) (Sengupta et al., 2025, 2023)

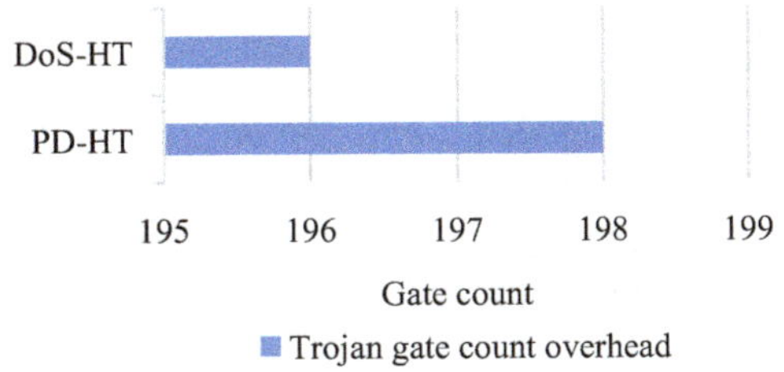

Figure 4.14 Trojan design area overhead (in terms of gate count) corresponding to convolutional layer CNN accelerator (Sengupta et al., 2025, 2024)

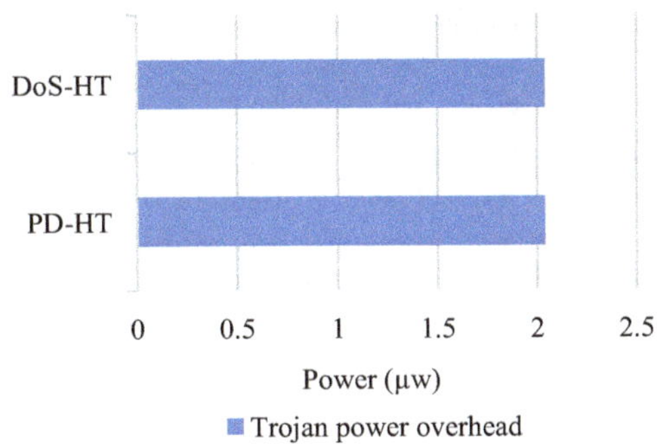

Figure 4.15 Trojan power overhead corresponding to convolutional layer CNN accelerator (Sengupta et al., 2025, 2024)

effects of Trojan injection on the design area (measured in terms of gate count) and the power overhead of the CNN accelerator. Trojan injection refers to the introduction of malicious alterations in the hardware design, typically to introduce vulnerabilities or alter performance characteristics in subtle ways. In this context, the figures demonstrate that after the Trojan infection, the CNN accelerator experiences a slight increase in both design area and power consumption compared to its baseline architecture. On average, the Trojan-infected design incurs only a modest increase in area, with a gate count rise of approximately 196 gates and power consumption escalation of around ~2 micro-watts (μw).

These case studies highlight the relatively minor impact of Trojan injection on the overall hardware overhead in terms of gate count and power, suggesting that such HLS Trojans can create devastating consequences on ML accelerator design. This insight is crucial for the design and security considerations of ML accelerators, where subtle changes can lead to profound shifts in performance and reliability, especially in critical applications such as autonomous HLS-based IP designing. It also indicates that advanced research in the domain of secure ML hardware accelerators is needed for the future.

4.7 Conclusion

This chapter has presented a detailed discussion on how ML hardware accelerators can be hacked by an attacker during their HLS design stages for injecting covert hardware Trojans. More explicitly, this chapter has elaborated on the HLS Trojan attack model, threat vector, triggering mechanism, and its payload on ML accelerators.

The chapter has lucidly presented details of some known ML accelerators, such as CNN convolutional layers and linear regression. The chapter also highlights the evasive capability of the HLS Trojan attack on ML accelerators, along with an analysis of standard metrics of power and area.

4.8 Questions and exercises

1. What is an ML accelerator?
2. What are the types of machine learning algorithms?
3. How is the impact of a PD-HT attack different from a DoS-HT attack on ML accelerators?
4. What are the different implementations of ML accelerators?
5. How do you justify the evasion capability of PD-HT and DoS-HT?
6. What are the different security vulnerabilities that exist during the ML accelerator design process?
7. What is NVDLA?
8. What is side channel leakage?
9. How does Trojan logic bypass cryptography cores?
10. How is the ML accelerator design process compromised?
11. What is NBTI stress?
12. What are the different stages of HLS?
13. How can the mux interconnect design be exploited for Trojan attacks?
14. What is the significance of the pre-defined/pre-loaded constant "c" in a Trojan attack?

References

A. Sengupta, A. Anshul, V. Chourasia, and N. Bhui, "Security vulnerability (Backdoor Trojan) during machine learning accelerator design phases," *IEEE IT Professional*, vol. 27, no. 1, pp. 65–72, 2025.

A. Anshul, and A. Sengupta, "A survey of high-level synthesis based hardware security approaches for reusable IP cores [Feature]," *IEEE Circuits and Systems Magazine*, vol. 23, no. 4, pp. 44–62, Fourthquarter 2023.

M. Arredondo-Velázquez, J. Diaz-Carmona, A. Barranco-Gutiérrez, and C. Torres-Huitzil, "Review of prominent strategies for mapping CNNs onto embedded systems," *IEEE Latin America Transactions*, vol. 18, no. 5, pp. 971–982, 2020.

A. Sengupta, and M. Rathor, "Obfuscated hardware accelerators for image processing filters—application specific and functionally reconfigurable processors," *IEEE Transactions on Consumer Electronics*, vol. 66, no. 4, pp. 386–395, 2020.

C. Jiang, D. Ojika, B. Patel, and H. Lam, "Optimized FPGA-based deep learning accelerator for sparse CNN using high bandwidth memory," in *Proceedings of the IEEE 29th Annual International Symposium on Field-Programmable Custom Computing Machines*., Orlando, FL, USA, 2021, pp. 157–164.

M. Xue, C. Gu, W. Liu, S. Yu, and M. O'Neill, "Ten years of hardware Trojans: a survey from the attacker's perspective," *IET Computers and Digital Techniques*, vol. 14, no. 6, pp. 231–246, 2020.

"Why you need HLS for machine learning accelerators." Siemens Digital Industries Software. Accessed: Oct. 2024. [Online]. Available: https://resources.sw.siemens.com/en-US/video-why-you-need-hls-for-machinelearning-accelerators

N. Gupta, A. Jati, and A. Chattopadhyay, "AI attacks AI: recovering neural network architecture from NVDLA using AI-assisted side channel attack," *Cryptology Archive*, Paper 2023/368, 2023. [Online]. Available: https://eprint.iacr.org/2023/368.pdf

D. Kachave, and A. Sengupta, "Digital processing core performance degradation due to hardware stress attacks," *IEEE Potentials*, vol. 38, no. 2, pp. 39–45, 2019.

A. Sengupta, R. Chaurasia, and M. Rathor, "HLS-based swarm intelligence driven optimized hardware IP core for linear regression-based machine learning," *IET Journal of Engineering*, vol. 2023, no. 8, 2023, Art. no. e12299.

A. Sengupta, A. Anshul, V. Chourasia, and N. Kumar, "M-HLS: malevolent high-level synthesis for watermarked hardware IPs," *IEEE Embedded Systems Letters*, vol. 16, no. 4, pp. 497–500, 2024.

A. Sengupta, and R. Chaurasia, "Secured convolutional layer IP core in convolutional neural network using facial biometric," *IEEE Transactions on Consumer Electronics*, vol. 68, no. 3, pp. 291–306, 2022.

M. Abderehman, R. Gupta, R. R. Theegala, and C. Karfa, "BLAST: belling the black-hat high-level synthesis tool," *IEEE Transactions on Computer-Aided Design of Integrated Circuits and Systems*, vol. 41, no. 11, pp. 3661–3672, 2022.

A. Sengupta, S. Bhadauria, and S. P. Mohanty, "TL-HLS: methodology for low cost hardware Trojan security aware scheduling with optimal loop unrolling factor during high level synthesis," *IEEE Transactions on Computer-Aided Design of Integrated Circuits and Systems*, vol. 36, no. 4, pp. 655–668, 2017.

Y. Huang, S. Bhunia, and P. Mishra, "Scalable test generation for Trojan detection using side channel analysis," *IEEE Transactions on Information Forensics and Security*, vol. 13, no. 11, pp. 2746–2760, 2018.

Y. Jin, and Y. Makris, "Hardware Trojan detection using path delay fingerprint," *Proceedings of the IEEE International Workshop on Hardware-Oriented Security and Trust*, 2008, pp. 51–57.

R. Yasaei, L. Chen, S.-Y. Yu, and M. A. A. Faruque, "Hardware Trojan detection using graph neural networks," *IEEE Transactions on Computer-Aided Design of Integrated Circuits and Systems*, early access, May 26, 2022.

M. Rathor, and A. Sengupta, "Revisiting black-hat HLS: a lightweight countermeasure to HLS-aided Trojan attack," *IEEE Embedded Systems Letters*, vol. 16, no. 2, pp. 170–173, 2024.

C. Pilato, K. Basu, F. Regazzoni, and R. Karri, "Black-hat high-level synthesis: myth or reality?" *IEEE Transactions on Very Large Scale Integration*, vol. 27, no. 4, pp. 913–926, 2019.

K. I. Gubbi *et al.*, "Securing AI hardware: challenges in detecting and mitigating hardware Trojans in ML accelerators," in *Proceedings of the IEEE 66th International Midwest Symposium on Circuits and Systems (MWSCAS), Tempe*, Tempe, AZ, USA, 2023, pp. 821–825.

Chapter 5

Power exhaustion attacks in HLS

Anirban Sengupta[1] *and Nabendu Bhui*[1]

High-level synthesis (HLS) has emerged as a standard design paradigm for designing digital intellectual property (IP) designs used in system-on-chips (SoCs) of several electronics/computing systems. However, due to the globalization of the design supply chain, HLS tools/frameworks can be considered untrustworthy. In this chapter, we discuss a very recent HLS Trojan attack that can be launched by an adversary (in the HLS tool vendor house). The chapter highlights this recent state-of-the-art HLS Trojan attack known as "TST-HT," along with its payload and triggering mechanism. The chapter also elaborates on the power drainage payload achieved by this HLS Trojan threat vector and compares it to other recent HLS Trojan attacks. The chapter summarizes the insertion strategy, evasion capability, and subtle nature of this TST-hardware Trojan (HT) threat vector. Finally, the chapter concludes with exercise questions for readers.

The rest of the chapter is structured as follows: Section 5.1 describes the introduction of the chapter; Section 5.1.1 discusses the salient features of this chapter; Section 5.2 provides a detailed discussion on state-of-the-art HLS Trojan attacks; Section 5.3 discusses the HLS Trojan tampering attack of IP designs; Section 5.3.1 describes the summary of the attack model; Section 5.3.2 discusses hardware Trojan (HT) insertion in the HLS flow; Section 5.3.3 describes the HLS Trojan framework; Section 5.4 discusses the implementation of Timer-Sensitive Trojan trigger-based automated logic-Transformation (TST-HT) insertion in HLS-generated IP design; Section 5.5 describes the analysis of case studies; and Section 5.6 concludes the chapter by summarizing its significance and implications.

5.1 Introduction

In the current system-on-chips (SoCs) of advanced multimedia, consumer electronics (CE), and medical systems, reusable intellectual property (IP) designs are fundamental building blocks (Rostami *et al.*, 2014; Sengupta, 2017a; Sengupta and Rathor, 2019; Sengupta and Rathor, 2020). Modern electronic, multimedia, and

[1]Deptartment of Computer Science and Engineering, Indian Institute of Technology Indore, India

healthcare technologies rely heavily on secure, application-specific processors and hardware, which are developed through electronic design automation (EDA) computer-aided design (CAD) workflows across multiple abstraction layers of IP design (Rostami *et al.*, 2014; Sengupta, 2017b; Sengupta and Bhui, 2025a; Sengupta and Bhui, 2025b; Bhui and Sengupta, 2025; Sengupta and Bhui, 2025c; Sengupta and Bhadauria, 2015). High-level synthesis (HLS) is widely used in the semiconductor industry as a standard EDA CAD methodology for creating such reusable IPs (Rostami *et al.*, 2014; Anshul and Sengupta, 2023; Sengupta *et al.*, 2025b, 2025c, 2023, 2025d; Sengupta and Bhadauria, 2016). The HLS framework includes several important stages, such as scheduling, functional unit allocation, register allocation, binding, designing the interconnectivity, building the finite state machine (FSM), and generating the datapath at the register-transfer level (RTL). Because of these roles, HLS-derived IPs are fundamental to highly computational/numerous applications in electronics, multimedia, and healthcare. Due to a higher design complexity, IP designers often use HLS to design IPs from a higher abstraction level of the integrated circuit (IC) design process, thus improving productivity and cutting down the development time.

However, scientific literature (Pilato *et al.*, 2019; Abderehman *et al.*, 2022; Sengupta *et al.*, 2017, 2024, 2023) demonstrates that the HLS design flow can introduce security vulnerabilities. A malicious HLS tool developer can take advantage of the automation process of the HLS tool/framework to secretly implant hardware Trojan (HT) backdoors, enabling attacks such as slowing system performance, altering functional behavior, leaking sensitive information, draining power, or undermining overall system security (Rathor and Sengupta, 2024; Jin *et al.*, 2008; Sengupta *et al.*, 2025c). Therefore, HLS tools/frameworks can be considered untrustworthy in nature. Practically every phase of the HLS design flow can be targeted, and existing attacks have already focused on stages such as the HLS module library, scheduling, allocation, mux-based interconnects, FSM generation, etc. Based on the specific type of Trojan payload injected/embedded during the HLS design flow, hardware Trojans can be categorized into several forms (Sengupta *et al.*, 2025c): functional hardware Trojan (F-HT) inserted through the HLS tool's library (as component) (Sengupta *et al.*, 2017); degradation hardware Trojan (DG-HT) implanted in scheduling phase (Pilato *et al.*, 2019); battery-exhaustion hardware Trojan (BE-HT) inserted in the FSM stage (Pilato *et al.*, 2019); downgrade-attack hardware Trojan (DA-HT) embedded via hidden HLS directives (Pilato *et al.*, 2019); performance-degradation hardware Trojan (PD-HT) implanted in the HLS scheduling phase (Sengupta *et al.*, 2024); and denial-of-service hardware Trojan (DoS-HT) introduced during interconnect design stage (Sengupta *et al.*, 2024), etc.

This chapter discusses an automated logic-transformation (LT)-based HLS hardware Trojan type that is secretly inserted during the datapath design stage of an HLS process for inducing a power drainage attack on IP designs (Sengupta *et al.*, 2025a). It behaves in an elusive manner, and it is difficult to detect, as it becomes active only when an attacker-specified, predetermined, specific, and rare-event triggering condition matches.

5.1.1 Salient features of this chapter

The salient features of this chapter are as follows:

- This chapter discusses a state-of-the-art (SOTA) HLS Trojan type, viz., Timer-Sensitive Trojan trigger-based automated logic-Transformation (TST-HT), that is capable of inducing a power drainage attack (Sengupta *et al.*, 2025a).
- This chapter elaborates on the automatic Trojan insertion strategy during the HLS datapath design stage.
- This TST-HT is capable of achieving significantly stronger power drainage than other SOTA techniques (Pilato *et al.*, 2019; Sengupta *et al.*, 2024), at zero latency (delay) overhead.

5.2 State-of-the-art on HLS Trojan attacks

To design and trigger an HLS-aided TST-HT, a combination of functional units (FUs), multiplexor (mux)-based switches, registers, magnitude comparator logic, and counters, along with a full datapath design stage (attacking stage), was introduced in the SOTA technique (Sengupta *et al.*, 2025a). It discussed a new method for Trojan insertion to adversely affect the power of a hardware IP design without affecting the performance (delay) and functionality. Moreover, no prior SOTA approaches have discussed injecting a timer-sensitive Trojan trigger-based automated logic-transformation (TST-HT) during the HLS process, especially by transforming the mathematical logic and exploiting the mux-based interconnect design stage. In the SOTA technique (Pilato *et al.*, 2019), the authors have explained a malicious HLS flow that can introduce a secret hardware trojan during the FSM design and scheduling phases for performance degradation and power exhaustion attacks. However, Pilato *et al.'s* (2019) technique does not introduce any TST-HT for draining the power, unlike the approach of Sengupta *et al.* (2025a). In addition, attacking a site/stage is also different in the approach of Pilato *et al.* (2019). Furthermore, Pilato *et al.* (2019) also do not discuss time-sensitive rare-event triggering, unlike the discussed approach. The approach by Sengupta *et al.* (2017) describes a functional Trojan framework in which a backdoor Trojan can be incorporated into an HLS module library. Here, authors have also introduced a detection technique using a dual modular redundancy (DMR) scheme with a distinct IP vendor allocation policy to detect functional Trojans. Nevertheless, Sengupta *et al.* (2017) do not address TST-HT-based HLS Trojans targeting power consumption, as introduced in the Sengupta *et al.* (2025a) framework, and are also limited to detecting only functional Trojans. In Abderehman *et al.* (2022), a Trojan detection approach is introduced that relies on C-to-RTL equivalence checking and is capable of detecting the DA-HT attack, BE-HT attack, and DG-HT attack discussed in Pilato *et al.* (2019). However, it cannot identify the HLS-aided TST-HT (Sengupta *et al.*, 2025a), since Abderehman *et al.* (2022) rely on having both the Trojan-infected RTL (generated via the Bambu tool) and the corresponding input C-code to perform equivalence checking. This is because the Bambu tool does not comprise transformation (LT)-based Trojan-infected RTL code for IP designs, as discussed in

Sengupta *et al.* (2025a). Moreover, the TST-HT-infected C-code is not available for analysis. The code-coverage analysis technique (Nahiyan and Tehranipoor, 2017) is used to detect suspicious signals in an RTL design and to generate additional test vectors for uncovered/untested parts/sections of the RTL design. However, this method is impractical for HLS-aided IP designs due to their high complexity. For example, in the approach (Sengupta *et al.*, 2025a), the Trojan-infected IP design may produce netlist complexity exceeding 20,000 lines of code. Applying code-coverage in such cases leads to excessive verification times. Additionally, as HLS-based IP designs become more complex, code-coverage analysis fails to achieve adequate coverage because the number of suspicious/malicious signals grows too large to manage effectively. In Sengupta *et al.* (2024), the authors have introduced a malicious HLS framework that covertly injects/embeds two Trojan-type PD-HT and DoS-HT during the mux-based interconnect design phase of the HLS process. Compared to Sengupta *et al.* (2025a), it does not consider HLS-supported TST-HT (embedded at the datapath design stage) to carry out a power drainage attack.

Moreover, some authors have discussed detection techniques such as Yasaei *et al.* (2025), Veeranna and Schafer (2017), Cui *et al.* (2014), and Rajmohan *et al.* (2020). However, none of these techniques is capable of detecting the discussed HLS-aided TST-HT. This is because Veeranna and Schafer (2017), Cui *et al.* (2014), and Rajmohan *et al.* (2020) can only detect functional Trojans where there is a change in the functionality or data due to the Trojan payload. However, the discussed HLS-aided TST-HT (Sengupta *et al.*, 2025a) does not change the functionality, data, or performance. Further, the discussed HLS-aided TST-HT only gets activated on a predefined timer-sensitive trigger when the attacker intends. This further exempts the possibility of detection using standard detection techniques. Finally, authors in Yasaei *et al.* (2025) discussed a detection technique by assuming that an attacker can insert HT directly in the Verilog RTL/netlist. However, in the discussed framework, HTs are inserted in the HLS datapath design stage (pre-RTL synthesis stage) of IP design. Moreover, Yasaei *et al.* (2025) can perform detection only using the Pyverilog toolkit for learning. However, it cannot handle complex Trojan-infected VHDL netlist codes generated in the discussed framework, and it exhibits weaker learning behavior for HLS-generated complex IP designs. This is because the discussed SOTA Trojan-infected VHDL netlist code (.vho) generates over several thousand lines of RTL code (Sengupta *et al.*, 2025a), which (Yasaei *et al.*, 2025) is unable to handle due to higher complexity. Therefore, the HLS-based TST-HT attack technique (Sengupta *et al.*, 2025a) is stealthy, more powerful in power degradation, and cannot be detected using SOTA hardware Trojan detection techniques.

5.3 Attack model: HLS Trojan tampering attack of IP designs

This section discusses the attack model for the HLS Trojan attack vector, the TST-HT insertion strategy, and its corresponding framework.

5.3.1 Summary of the attack model

SOTA technique/research (Pilato *et al.*, 2019; Abderehman *et al.*, 2022; Sengupta *et al.*, 2017) has shown that the HLS framework/tool (due to its untrustworthy nature) may exhibit security vulnerabilities across several phases of the IP design flow, including the module library, scheduling stage, interconnect stage, and controller/finite-state-machine stage. These vulnerabilities arise mainly because (i) these may be unreliable, untrustworthy and compromised and might potentially include backdoor Trojans without any knowledge of the original IP designer/vendor (who uses this HLS tool); (ii) malicious logic can be secretly injected/embedded into intermediate HLS stages, where it can stay quiescent/inactive during normal/regular operation and activate only under adversarial trigger condition; and (iii) compromised HLS tools/frameworks (containing embedded Trojans) are part of a national-level toolkit, where an HLS tool developed/engineered in one country could be integrated into the design workflow of another country. Thus, an attacker/adversary with privileged access to the HLS tool development process can stealthily insert/embed a backdoor Trojan into any one of the stages of the HLS design process.

This chapter discusses an SOTA HLS Trojan type "TST-HT," that is secretly inserted during the HLS datapath design stage for inducing a power drainage attack on IP designs (Sengupta *et al.*, 2025a). This covert insertion of the TST-HT in the datapath during the HLS flow is performed in such a way that it impacts the power drainage significantly as a Trojan payload but without disturbing the functionality, output data, or performance (latency/delay). Since HLS-generated RTL IP designs are significantly larger and more convoluted than their high-level specification equivalents, any hidden Trojan logic is significantly challenging to detect. Further, due to its stealthy nature, owing to the time-sensitive rare-event triggering, as well as being insensitive to functional change/output data/performance, such a class of Trojans [like the one discussed in this chapter (Sengupta *et al.*, 2025a)] becomes hard to detect.

5.3.2 HT insertion in the HLS flow: timer-sensitive Trojan trigger-based automated logic-transformation (TST-HT)

In the HLS datapath design stage, the structure and connectivity of the IP designs are defined, including how components are interconnected using switches or multiplexers. An attacker or malicious designer working in the HLS tool development environment can secretly alter the logic of an operation by introducing an automated logic-transformation HT into the HLS datapath design (associated with the data flow graph) without modifying/changing the functionality, output, or performance. The TST-HT insertion technique (Sengupta *et al.*, 2025a) involves a covert alteration of the HLS datapath architecture that activates a power drainage payload only when a specific, timer-driven rare-event trigger condition is fulfilled. This hardware Trojan logic signal only gets activated when a predefined time has elapsed, i.e., the modulus (mod) up-counter value reaches a constant value that is already predefined and preloaded in an in-built storage/register. Figure 5.1

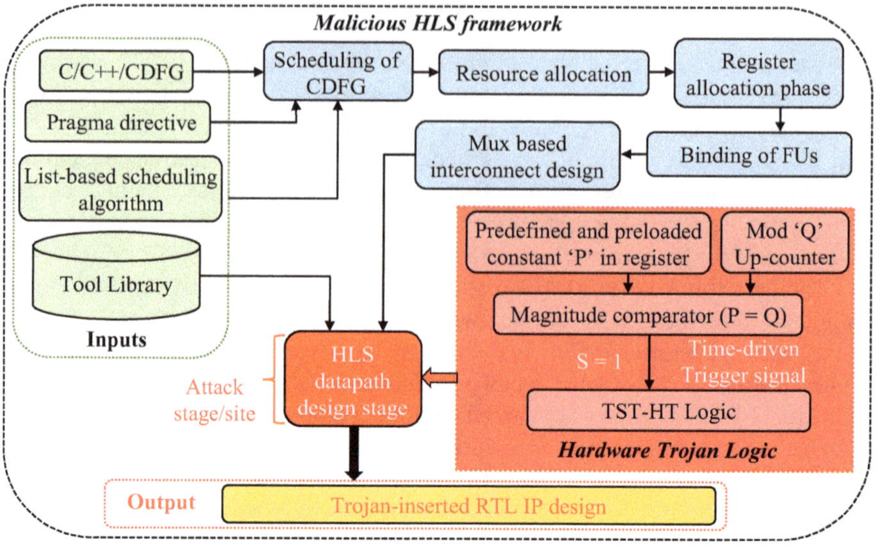

Figure 5.1 TST-HT insertion stage in the HLS design flow

illustrates the insertion/attack stage of the TST-HT in the HLS framework. As shown in Figure 5.1, the primary inputs of the HLS framework comprise pragma directives/resource constraints, the C/C++/control data flow graph (CDFG) of the intended hardware application, the HLS tool library, and the list-based scheduling algorithm. The HLS tool library comprises a collection of predefined blocks/components used to design the datapath of the hardware IP design at RTL. Further, it passes through the HLS process/flow, like scheduling of CDFG, resource allocation, register allocation, FU binding, and mux-based interconnect design, to generate the HLS datapath. As depicted in Figure 5.1, the TST-HT is activated (Trojan-driven trigger signal *at* $S = 1$) when the mod "Q" up-counter reaches the predefined and preloaded constant "P" under a specific rare-event condition. Under all other conditions, the Trojan remains dormant, and it is completely untraceable as functionality/output is not affected. As this automated logic transformation-based hardware Trojan activates only during a specific rare-event condition, without altering functionality, output data, or performance (delay), while inducing unwarranted power drainage, detecting this type of hardware Trojan is extremely difficult and challenging.

5.3.3 HLS Trojan framework

To illustrate the TST-HT, the authors in Sengupta *et al.* (2025a) have adopted a CDFG from Sengupta *et al.* (2018). As depicted in Figure 5.2, the scheduled data flow graph (SDFG) consists of nine operations (opn 1–9); three resources as IP vendors specified constraints (two multipliers and one adder); ten inputs: $i_1, i_2, i_3, i_4, \ldots, i_{10}$; one output (O/P): v_8; and scheduled in control steps *CS0–CS5*. During

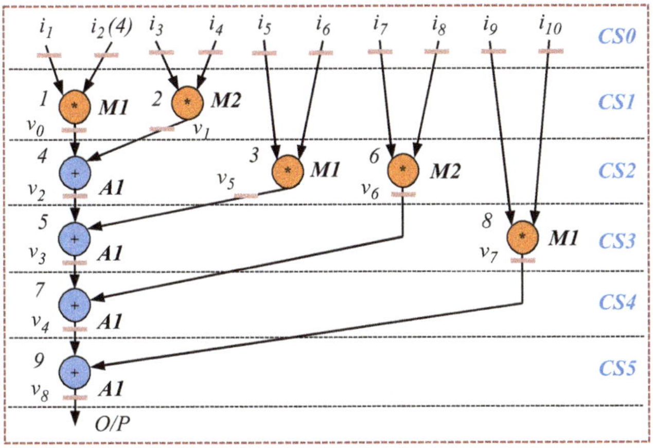

FUNCTIONAL UNIT WITH CORRESPONDING REQUIRED MUX AND DEMUX FOR TEST CASE SDFG

FU	MUX	DEMUX
M1	4:1 → 2	1:4 → 1
M2	2:1 → 2	1:2 → 1
A1	4:1 → 2	1:4 → 1

Figure 5.2 Scheduled data flow graph (SDFG) of test case control data flow graph (CDFG) with 1A(+) and 2M(*)

the HLS datapath design process, the required mux-based switching connections corresponding to each resource are established. As shown in Figure 5.2 (test case SDFG), designing the HLS datapath requires a total of four 4:1 multiplexers (A1→2, M1→2), two 2:1 multiplexers (M2→2), two 1:4 demultiplexers (A1→1, M1→1), a single 1:2 demultiplexer (M2→1), and 19 registers. A malicious designer or attacker can covertly influence the logic of a particular operation (opn 1 in Figure 5.2) by injecting TST-HT into the HLS datapath, without making any changes to the original output or affecting functionality. Further, to make the TST-HT payload operational (causing excessive power drainage), an attacker initiates a timer-controlled rare trigger while the HLS datapath is active. Moreover, the attacker incorporates the TST-HT logic directly into the datapath during the HLS tool development phase. This leads to an automatic modification of the logic, as illustrated in Figure 5.3. Actually, the execution of TST-HT leads to automated conversion of the multiplication operation to its equivalent functional logic in the form of successive additions (shown in Figure 5.3). The attacker implements this HLS Trojan logic in the datapath design stage. After the TST-HT embedding, the datapath exhibits a different interconnect logic configuration. For example (from test case SDFG), operation 1 (M1) is replaced by operation 1 (A1) and operation 1′ (A1) to form a functionally equivalent logic. Therefore, the number of FU is updated accordingly. In the Trojan-free SDFG

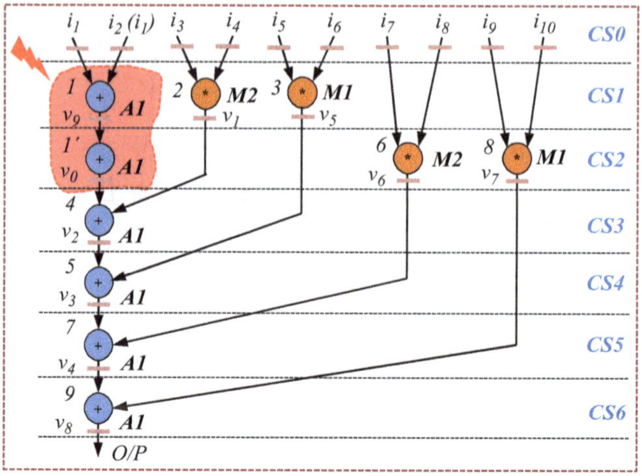

FUNCTIONAL UNIT WITH CORRESPONDING REQUIRED MUX AND DEMUX FOR TEST CASE SDFG AFTER TROJAN INDUCING

FU	MUX	DEMUX
M1	2:1 → 2	1:2 → 1
M2	2:1 → 2	1:2 → 1
A1	8:1 → 2	1:8 → 1

Figure 5.3 Trojan-induced tampered SDFG of test case CDFG

(Figure 5.2), the multiplier M1 (i.e., opn 1) has inputs i_1 and $i_2 = 4$, which generates output v_0. In contrast, in the TST-HT inserted SDFG (Figure 5.3), adder A1 (opn 1) receives inputs i_1 and $i_2 = i_1$, which generates output v_9; further, adder A1 (opn 1′) takes v_9 as inputs and finally generates output v_0. This ensures that the overall functional output remains unchanged. As depicted in Figure 5.4, the partial view of the IP datapath reflects the corresponding multiplexers and demultiplexers associated with these FUs after TST-HT insertion. During Trojan-free execution, the SDFG (Figure 5.2) contains four addition operations with respective input pairs ($<v_0, v_1>$, $<v_2, v_5>$, $<v_3, v_6>$, $<v_4, v_7>$) and corresponding outputs (v_2, v_3, v_4, v_8); all of these inputs and outputs are connected to the adder (+) resource via the mux-demux switching connection/logic. During Trojan activation, it converts into a total of two addition (+) operations in place of one multiplication (*) operation. The SDFG (Figure 5.3) contains six addition operations with respective input pairs ($<i_1, i_2>$, $<v_9, v_9>$, $<v_0, v_1>$, $<v_2, v_5>$, $<v_3, v_6>$, $<v_4, v_7>$) and corresponding outputs ($v_9, v_0, v_2, v_3, v_4, v_8$); all of these inputs and outputs are connected to the adder (+) resource via the mux-demux switching connection/logic. As shown in the figure (Figure 5.4), a rogue HLS tool developer introduces the TST-HT as a malicious alteration to the HLS datapath, triggering a power-draining payload under a timer-based rare-event condition (@ $S = 1$). Here, red-colored components indicate the TST-HT logic, while other

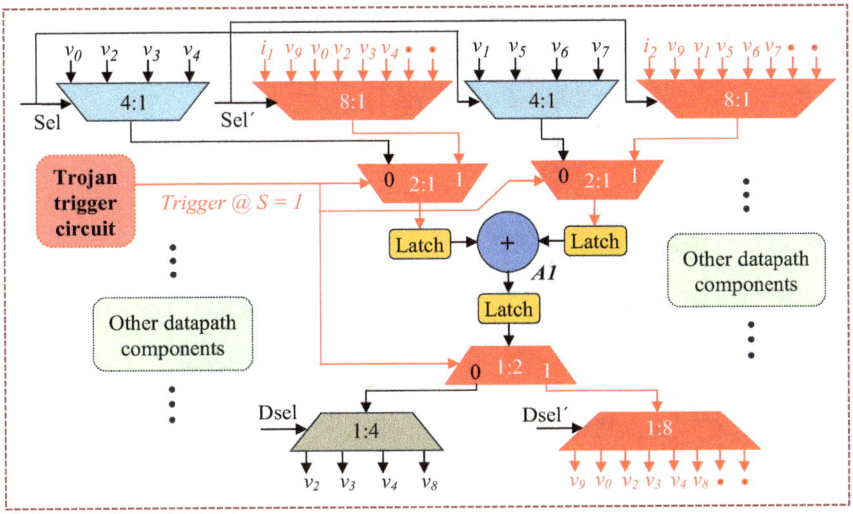

Figure 5.4 Partial view of TST-HT-induced datapath for test case SDFG

components/logic are part of the regular HLS datapath design. Even on activation of the Trojan payload, the TST-HT is not detectable as its functionality and performance (delay) are not affected. Subsequently, Figure 5.5 shows the complete HLS datapath IP (for test case SDFG), where red-colored modules identify the embedded TST-HT logic contributing to significant power drainage/dissipation.

5.4 Implementation of TST-HT insertion in HLS-generated IP design

The implementation of TST-HT insertion is demonstrated on test case SDFG (as an example). A partial view of the implemented TST-HT inserted datapath design (in VHDL) is shown in Figure 5.6. This partial figure (Figure 5.6) shows the port mappings and corresponding signals for some connecting components. The TST-HT infected/embedded datapath post RTL synthesis in the Intel Quartus tool (within Cyclone IV GX FPGA) is shown in Table 5.1. It shows that the total logical elements and total combinational functions utilized are <1% in Cyclone IV GX. Further, Figures 5.7 and 5.8 show the partial view of the synthesized netlist file of the discussed TST-HT inserted datapath. The netlist file (.vho) of the discussed rate-event-triggered TST-HT-infected datapath comprises ~22,186 lines of code (post-HLS and RTL synthesis). Therefore, it becomes literally impossible to trace the discussed TST-HT in the complex HLS-generated RTL IP designs using code-coverage detection analysis (Nahiyan and Tehranipoor, 2017).

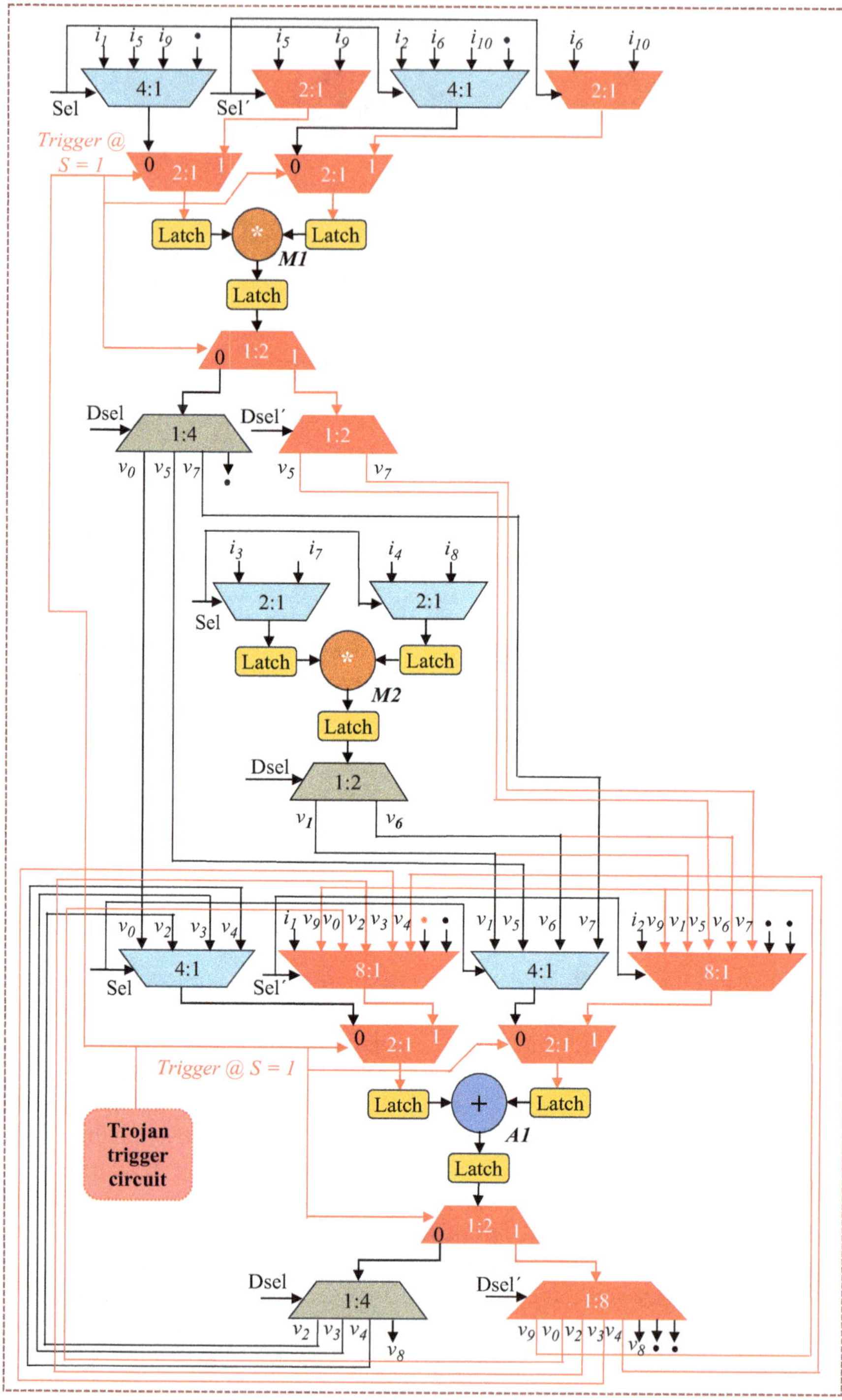

Figure 5.5 Full view of the TST-HT-induced datapath

```
Q1:mux_2to1 PORT MAP(A2=>IN_T1,B2=>IN_T2,MS2=>TS1_1,MZ2=>T1);
Q2:mux_2to1 PORT MAP(A2=>IN_T3,B2=>IN_T4,MS2=>TS1_1,MZ2=>T2);
Q3:mux_2to1 PORT MAP(A2=>S1,B2=>T1,MS2=>S1_2,MZ2=>T3);
Q4:mux_2to1 PORT MAP(A2=>S2,B2=>T2,MS2=>S1_2,MZ2=>T4);
Q5:demux_1to2 PORT MAP(P2=>S7,Q2=>T5,DS2=>S1_2,Z2=>S6);
Q6:demux_1to2 PORT MAP(P2=>V5_1,Q2=>V7_1,DS2=>S1_3,Z2=>T5);
Q7:mux_8to1 PORT MAP(A8=>IN_T5,B8=>V9_1,C8=>V0_1,D8=>V2_1,E8=>V3_1,F8=>V4_1,G8=>V_1,H8=>V_2,MS8=>S3_1,MZ8=>T6);
Q8:mux_8to1 PORT MAP(A8=>IN_T6,B8=>V9_1,C8=>V1_1,D8=>V5_1,E8=>V6_1,F8=>V7_1,G8=>V_3,H8=>V_4,MS8=>S3_1,MZ8=>T7);
Q9:mux_2to1 PORT MAP(A2=>S14,B2=>T6,MS2=>S1_2,MZ2=>T8);
Q10:mux_2to1 PORT MAP(A2=>S15,B2=>T7,MS2=>S1_2,MZ2=>T9);
Q11:demux_1to2 PORT MAP(P2=>S20,Q2=>T10,DS2=>S1_2,Z2=>S19);
Q12:demux_1to8 PORT MAP(P0=>V9_1,P1=>V0_1,P2=>V2_1,P3=>V3_1,P4=>V4_1,P5=>V_81,P6=>V_5,P7=>V_6,DS8=>D3_1,DZ8=>T10);
----------------------------MUL_1---------------------------
Q13:mux_4to1 PORT MAP(A4=>IN1,B4=>IN5,C4=>IN9,D4=>V_7,MS4=>S2_1,MZ4=>S1);
Q14:mux_4to1 PORT MAP(A4=>IN2,B4=>IN6,C4=>IN10,D4=>V_9,MS4=>S2_1,MZ4=>S2);
Q15:latch1 PORT MAP(L1=>T3,LEN1=>W1,LZ1=>S3);
Q16:latch1 PORT MAP(L1=>T4,LEN1=>W2,LZ1=>S4);
Q17:mul PORT MAP(mA=>S3,mB=>S4,mZ=>S5);
Q18:latch1 PORT MAP(L1=>S5,LEN1=>W3,LZ1=>S6);
Q19:demux_1to4 PORT MAP(P4=>V0,Q4=>V5,R4=>V7,T4=>V_7,DS4=>D2_1,DZ4=>S7);
--------------------------------MUL_2---------------------
Q20:mux_2to1 PORT MAP(A2=>IN3,B2=>IN7,MS2=>S1_1,MZ2=>S8);
Q21:mux_2to1 PORT MAP(A2=>IN4,B2=>IN8,MS2=>S1_1,MZ2=>S9);
```

Figure 5.6 Partial view of the VHDL code of the implemented TST-HT-inserted datapath design

Table 5.1 FPGA synthesis results

Device utilization summary of the Trojan-infected datapath of CDFG	
Family	**Cyclone IV GX**
Total logic elements	<1%
Total combinational functions	<1%

```
-- pragma translate_on
PORT MAP (
        dataa => \Q11|Q2\(13),
        datab => \D3_1[1]~input_o\,
        datac => \D3_1[2]~input_o\,
        datad => \D3_1[0]~input_o\,
        combout => \Q12|P5[13]~13_combout\);

-- Location: LCCOMB_X60_Y26_N30
\Q12|P5[14]~14\ : cycloneiv_lcell_comb
-- Equation(s):
-- \Q12|P5[14]~14_combout\ = (\D3_1[0]~input_o\ & (\D3_1[2]~input_o\ & (\Q11|Q2\(14) & !\D3_1[1]~input_o\)))

-- pragma translate_off
GENERIC MAP (
        lut_mask => "0000000010000000",
        sum_lutc_input => "datac")
-- pragma translate_on
PORT MAP (
        dataa => \D3_1[0]~input_o\,
        datab => \D3_1[2]~input_o\,
        datac => \Q11|Q2\(14),
        datad => \D3_1[1]~input_o\,
        combout => \Q12|P5[14]~14_combout\);

-- Location: LCCOMB_X59_Y22_N16
\Q12|P5[15]~15\ : cycloneiv_lcell_comb
-- Equation(s):
-- \Q12|P5[15]~15_combout\ = (\Q11|Q2\(15) & (\D3_1[0]~input_o\ & (!\D3_1[1]~input_o\ & \D3_1[2]~input_o\)))

-- pragma translate_off
GENERIC MAP (
```

Figure 5.7 Partial views of the netlist file of the proposed TST-HT-inserted datapath

```
-- pragma translate_off
GENERIC MAP (
        lut_mask => "1111101000001010",
        sum_lutc_input => "datac")
-- pragma translate_on
PORT MAP (
        dataa => \Q29|LZ1\(15),
        datac => \W7~inputclkctrl_outclk\,
        datad => \Q9|MZ2[15]~71_combout\,
        combout => \Q29|LZ1\(15));

-- Location: LCCOMB_X59_Y26_N30
\Q31|aZ[15]~30\ : cycloneiv_lcell_comb
-- Equation(s):
-- \Q31|aZ[15]~30_combout\ = \Q30|LZ1\(15) $ (\Q31|aZ[14]~29\ $ (\Q29|LZ1\(15)))

-- pragma translate_off
GENERIC MAP (
        lut_mask => "1010010101011010",
        sum_lutc_input => "cin")
-- pragma translate_on
PORT MAP (
        dataa => \Q30|LZ1\(15),
        datad => \Q29|LZ1\(15),
        cin => \Q31|aZ[14]~29\,|
        combout => \Q31|aZ[15]~30_combout\);

-- Location: LCCOMB_X60_Y29_N24
\Q32|LZ1[15]\ : cycloneiv_lcell_comb
```

Figure 5.8 Another partial view of the netlist of the proposed TST-HT inserted datapath

5.5 Analysis of case studies

The performance of the TST-HT integrated HLS datapath IP design is assessed using Silvaco's 15 nm OpenCell library (Silvaco's Open Cell Library, 2025).

The TST-HT framework inserted during HLS is capable of producing significant power drainage, at 0% latency overhead as reported in Tables 5.2 and 5.3 (Sengupta *et al.*, 2025a). As shown in Table 5.2, the TST-HT framework incurs a power overhead of 19.46% for the test case CDFG, 25.79% for the sharpening filter, 34% for the edge detection filter, 18.57% for the blur filter, and 44.86% for the cardiac pacemaker filter. The table also indicates the IP datapath components utilized during the HLS datapath design before Trojan insertion (Trojan-free) and after TST-Trojan logic insertion. On average, the TST-HT logic is able to induce power drainage of ~28.54%. Further, Table 5.3 shows the delay (latency) overhead incurred due to TST-HT insertion in the HLS flow. As evident from Table 5.3, the Trojan-induced datapath IP produces a similar delay as the Trojan-free datapath IP. In some cases, due to logic transformation-based Trojan logic insertion, the delay post TST-HT insertion is less than the Trojan-free design. This is because, for some designs such as the test case CDFG and the edge detection filter, the number of multiplications reduces to a succession of additions post-logic transformation. This causes the latency of the schedule to reduce compared to the latency of the Trojan-free design. Therefore, as evident, for all IP designs, the discussed TST-HT HLS framework causes significant power drainage at zero latency (delay) overhead.

The TST-HT HLS framework is also compared with Black-HAT (Pilato *et al.*, 2019) and M-HLS (Sengupta *et al.*, 2024) in terms of power drainage achieved. As

Table 5.2 Power drainage achieved by the TST-HT-infected IP datapath for different benchmarks

Benchmarks	Trojan-free datapath IP (in dormant state)		Trojan-induced datapath IP (post-activation)		Power drainage achieved by the TST-HT framework
	Datapath components	Power (μW)	Datapath components	Power (μW)	
Test case	Reg. → 19 MUX 2:1 → 2 DEMUX 1:2 → 1 MUX 4:1 → 4 DEMUX 1:4 → 2 Adder (A) → 1 Multiplier (M) → 2	36.38	Reg. → 20 MUX 2:1 → 8 DEMUX 1:2 → 4 MUX 8:1 → 2 DEMUX 1:8 → 1 Adder (A) → 1 Multiplier (M) → 2	43.46	7.08 μW
Blur filter	Reg. → 38 MUX 2:1 → 2 DEMUX 1:2 → 1 MUX 8:1 → 4 DEMUX 1:8 → 2 Adder (A) → 2 Multiplier (M) → 1	68.65	Reg. → 39 MUX 2:1 → 5 DEMUX 1:2 → 3 MUX 8:1 → 2 DEMUX 1:8 → 1 MUX 16:1 → 2 DEMUX 1:16 → 1 Adder (A) → 2 Multiplier (M) → 1	81.40	12.75 μW
Sharpening filter	Reg. → 21 MUX 2:1 → 2 DEMUX 1:2 → 1 MUX 4:1 → 4 DEMUX 1:4 → 2 Adder (A) → 2 Multiplier (M) → 1	37.28	Reg. → 22 MUX 2:1 → 5 DEMUX 1:2 → 3 MUX 4:1 → 2 DEMUX 1:4 → 1 MUX 8:1 → 2 DEMUX 1:8 → 1 Adder (A) → 2 Multiplier (M) → 1	46.89	9.61 μW

(Continues)

Table 5.2 (Continued)

Benchmarks	**Trojan-free datapath IP (in dormant state)**		**Trojan-induced datapath IP (post-activation)**		**Power drainage achieved by the TST-HT framework**
	Datapath components	**Power (μW)**	**Datapath components**	**Power (μW)**	
Cardiac pacemaker filter	Reg. → 14 MUX 2:1 → 2 DEMUX 1:2 → 1 MUX 4:1 → 2 DEMUX 1:4 → 1 Adder (A) → 1 Multiplier (M) → 1	24.33	Reg. → 15 MUX 2:1 → 8 DEMUX 1:2 → 3 MUX 8:1 → 2 DEMUX 1:8 → 1 Adder (A) → 1 Multiplier (M) → 1	35.24	10.91 μW
Edge detection filter	Reg. → 26 MUX 4:1 → 2 DEMUX 1:4 → 1 MUX 8:1 → 2 DEMUX 1:8 → 1 Adder (A) → 1 Multiplier (M) → 1	47.00	Reg. → 27 MUX 2:1 → 4 DEMUX 1:2 → 2 MUX 4:1 → 2 DEMUX 1:4 → 1 MUX 16:1 → 2 DEMUX 1:16 → 1 Adder (A) → 1 Multiplier (M) → 1	62.98	15.98 μW

Table 5.3 Latency/delay overhead of the TST-HT-infected IP datapath w. r. t. Trojan-free design for different benchmarks

Benchmarks	**Trojan-free datapath IP (ps)**	**Trojan-induced datapath IP (ps)**	**Delay overhead**	**Delay overhead**
Test case	927.40	794.91	−14.29%	0.00%
Blur filter	1059.88	1059.88	0.00%	
Sharpening filter	794.91	794.91	0.00%	
Cardiac pacemaker filter	728.67	728.67	0.00%	
Edge detection filter	1391.10	1324.86	−4.76%	

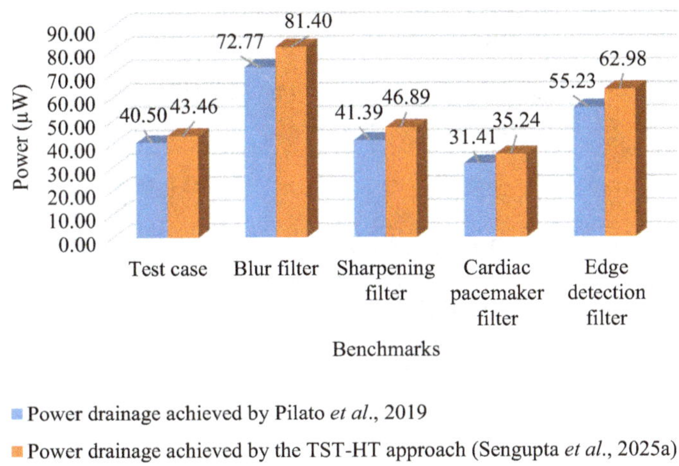

Figure 5.9 Power drainage comparison of the TST-HT approach with Pilato et al. (2019)

indicated in Figures 5.9 and 5.10, this TST-hardware Trojan framework exhibits higher power consumption than the approaches in Pilato *et al.* (2019) and Sengupta *et al.* (2024). Compared to Pilato *et al.* (2019), this TST-HT method increases power drainage by approximately 7.31% to 14.03%, while the improvement over Sengupta *et al.* (2024) ranges from 11.69% to 31.20%. The enhancement in the TST-HT framework compared to Pilato *et al.*'s (2019) and Sengupta *et al.*'s (2024) frameworks is because of the logic transformation induced by the discussed Trojan logic. On the contrary, Pilato *et al.* (2019) only exploit fake operation insertion during the scheduling phase, which causes lower power exhaustion than the TST-HT approach. Further, Sengupta *et al.* (2024) only exploit the inverter chain on the unutilized mux-ports during the mux-interconnect design stage, which also causes lower power exhaustion than the TST-HT approach.

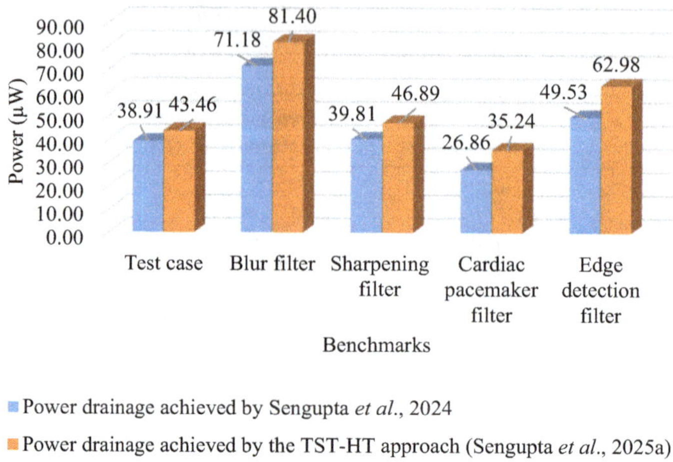

Figure 5.10 Power drainage comparison of the TST-HT approach with Sengupta et al. (2024)

As shown in Table 5.4 (Sengupta *et al.*, 2025a), the TST-hardware Trojan framework demonstrates its ability to evade several SOTA detection approaches, such as Abderehman *et al.* (2022), Sengupta *et al.* (2017), Yasaei *et al.* (2025), Cui *et al.* (2014), and Rajmohan *et al.* (2020). The detection approach based on C-to-RTL equivalence checking (Abderehman *et al.*, 2022) involves generating an FSM with a datapath and validating it against the corresponding C-code. In the TST-HT framework, Trojan components are integrated into the HLS datapath in a manner that preserves functional correctness. The method described in Abderehman *et al.* (2022) requires access to both the Trojan-compromised RTL (produced using the Bambu tool) and the original C source code to initiate equivalence checking. However, the Bambu tool does not support RTL IP designs infected with TST-HT Trojans as presented in this discussed approach (Sengupta *et al.*, 2025a). In addition, the TST-HT injected C source code is also needed for such evaluation, which is unavailable to analyze, and TST-HT triggers only on rare-event conditions, which makes it elusive by nature. The TL-HLS detection method (Sengupta *et al.*, 2017) incorporates DMR along with a distinct IP vendor allocation strategy, and it is limited to identifying functional Trojans introduced at the HLS level. Therefore, it cannot detect the TST-HT as it employs a power drainage payload without changing any functionality. Further, the graph neural network (GNN)-based detection method (Yasaei *et al.*, 2025) depends on Pyverilog, which cannot process VHDL RTL designs and shows reduced learning effectiveness when applied to complex IP designs generated through HLS. Similarly, Cui *et al.* (2014) are only capable of handling functional digital Trojans that can affect the logic values (i.e., can change the output of the circuit) using detection/recovery logic with a minimum of four distinct IP vendor allocation

Table 5.4 Trojan evasion capability against known detection techniques

Detection approach	**Trojan status**	**Remarks**
TL-HLS (Sengupta *et al.*, 2017)	Not detected	TL-HLS detection technique with DMR design is only able to handle HLS functional Trojans (unable to detect proposed LT-HT with power drainage payload)
GNN (Yasaei *et al.*, 2025)	Not detected	GNN-based detection technique uses Pyverilog, which is incapable of handling VHDL codes/RTL circuits and generates weaker learning behavior for complex IP designs
C to RTL equivalence checking (Abderehman *et al.*, 2022)	Not detected	As this equivalence checking relies on the creation of a finite state machine with a datapath (RTL) for comparing with C-code, hence, it is only capable of detecting BE-based HLS Trojans and DA HLS Trojans
Property checking for BIPs (Veeranna and Schafer, 2017)	Undetectable	Only capable of detecting functional HW Trojans in 3PBIPs, where the modeling of the HW Trojan is done using ternary operators. Needs a property checker feature in the HLS tool for detection. Only applicable for the HLS tool from CyberWorkBench. Does not work for other HLS tools/frameworks. Cannot detect power drainage HT like proposed LT-HT
HLS Trojan detection and recovery (Cui *et al.*, 2014)	Undetectable	It is only capable of handling functional digital Trojans that can affect the logic values (i.e., can change the output of the circuit) using detection/recovery logic with a minimum of four distinct IP vendor allocation policies. Cannot detect power drainage HT like proposed LT-HT
3PIP defence (Rajmohan *et al.*, 2020)	Undetectable	It is only capable of handling functional Trojans (i.e., Trojans that change the output of the circuit) using redundancy logic with a minimum of three distinct IP vendor allocation policies. Cannot detect power drainage HT like proposed LT-HT

(Note: DMR—Dual modular redundancy from Sengupta *et al.* (2017); BIPs: behavioral IPs; 3PIP: Third-party IP)

policies. It cannot detect power drainage HT like TST-HT. Further, the method (Rajmohan *et al.*, 2020) is limited to detecting the functional Trojans using redundancy logic across a minimum of three different IP vendors and cannot detect power drainage hardware Trojans like TST-HT. Furthermore, Table 5.5

Table 5.5 Trojan evasion capability against known SOTA detection techniques

Detection approach	Features in IP (baseline)—Design A (U^{OG})	Features in IP (with TST-HT)—Design B (U^{DP})	Trojan status
Sengupta *et al.* (2017)	Baseline design output = 8 (Note: a sample value is reported based on a specific test pattern provided. A large, exhaustive set of test patterns was tested) Sample test vectors for U^{OG}: $\{i_1 = i_3 = \ldots = i_{10} = 1\}$, and $i_2 = 4$	Trojan-infected design output = 8 (Note: a sample value is reported based on a specific test pattern provided. A large, exhaustive set of test patterns was tested) Sample test vectors for U^{DP}: $\{i_1 = i_3 = \ldots = i_{10} = 1\}$, and $i_2 = i_1$	✕
Rajmohan *et al.* (2020)	Baseline design output = 24 (Note: a sample value is reported based on a specific test pattern provided. A large, exhaustive set of test patterns was tested) Sample test vectors for U^{OG}: $\{i_1 = i_3 = \ldots = i_{10} = 2\}$, and $i_2 = 4$	Trojan-infected design output = 24 (Note: a sample value is reported based on a specific test pattern provided. A large, exhaustive set of test patterns was tested) Sample test vectors for U^{DP}: $\{i_1 = i_3 = \ldots = i_{10} = 2\}$, and $i_2 = i_1$	✕
Cui *et al.* (2014)	U^{OG}: allocated to 3PIP vendor 1 U^{OG} o/p = 48 Sample test vectors for U^{OG}: $\{i_1 = i_3 = \ldots = i_{10} = 3\}$, and $i_2 = 4$	U^{DP}: allocated to 3PIP vendor 2 U^{DP} o/p = 48 Sample test vectors for U^{DP}: $\{i_1 = i_3 = \ldots = i_{10} = 3\}$, and $i_2 = i_1$	✕

(Note: UOG and UDP indicate original and duplicate units from Sengupta *et al.* (2017); and "✕" means "Not detected")

(Sengupta *et al.*, 2025a) illustrates the evaluation demonstrating the HLS-based TST-HT framework's ability to evade SOTA detection methods. As shown in Table 5.5, detection techniques (Sengupta *et al.*, 2017; Cui *et al.*, 2014; and Rajmohan *et al.*, 2020) are inadequate to detect the discussed TST-hardware Trojan.

Further, the TST-HT embedded HLS datapath is complex and consists of more than a few thousand lines of RTL code, which makes the detection process extremely difficult and challenging. For example, the netlist file (.vho) of the rare-event triggered TST-HT infected datapath for test case CDFG comprises ~22,186 lines of code (post-HLS and RTL synthesis). Therefore, it becomes very tough to trace the TST-HT in the complex HLS-generated RTL IP designs. Further, such stealthy Trojans remain dormant during normal execution and only become activated on timer-sensitive, predefined rare-event conditions, without affecting the functionality/output.

5.6 Conclusion

This chapter provides an elaborate discussion on timer-sensitive Trojan trigger-based automated logic-transformation (TST-HT) and its evasion capability. This TST-hardware Trojan payload is injected/embedded in HLS-based DSP IP designs (in the datapath design stage) and is capable of inducing a power drainage attack. This TST-HT is capable of achieving significantly stronger power drainage than other SOTA techniques at zero latency (delay) overhead. Finally, the justifications of the elusive nature of TST-HT against standard detection techniques are also discussed. This chapter highlights the powerful nature of such timer-sensitive Trojan trigger-based automated logic transformation in the context of HLS IP designs.

5.7 Questions and exercises

1. What are the different insertion sites for HLS Trojans?
2. What is TST-HT?
3. How is TST-HT different from F-HT?
4. How is TST-HT different from PD-HT?
5. How is TST-HT different from DoS-HT?
6. How do you determine the time-sensitive nature of TST-HT?
7. How is PD-HT different from DoS-HT?
8. What kind of Trojan can be detected using SOTA detection techniques? Is it possible to detect TST-HT using existing SOTA detection techniques?
9. Why is C-to-RTL checking not capable of detecting TST-HT?
10. How is power drainage estimated for TST-HT-inserted IP designs?
11. Why is the code-coverage analysis technique not capable of detecting TST-HT?

References

A. Sengupta, N. Bhui, and V. Chourasia, "HLS Trojan Tampering Attack of IP Designs for Power Drainage using Timer-Sensitive Trojan Trigger based Automated Logic-Transformation," *IEEE Embedded Systems Letters*, 2025a, doi:10.1109/LES.2026.3656894.

C. Pilato, K. Basu, F. Regazzoni, and R. Karri, "Black-Hat high-level synthesis: myth or reality?," *IEEE Transactions on Very Large Scale Integration (VLSI) Systems*, vol. 27, no. 4, pp. 913–926, 2019, doi:10.1109/TVLSI.2018.2884742.

M. Abderehman, R. Gupta, R. R. Theegala, and C. Karfa, “BLAST: belling the Black-Hat high-level synthesis tool,” *IEEE Transactions on Computer-Aided Design of Integrated Circuits and Systems*, vol. 41, no. 11, pp. 3661–3672, 2022, doi:10.1109/TCAD.2022.3200513.

A. Sengupta, S. Bhadauria, and S. P. Mohanty, “TL-HLS: methodology for low cost hardware Trojan security aware scheduling with optimal loop unrolling factor during high level synthesis,” *IEEE Transactions on Computer-Aided Design of Integrated Circuits and Systems*, vol. 36, no. 4, pp. 655–668, 2017, doi:10.1109/TCAD.2016.2597232.

A. Sengupta, S. P. Mohanty, F. Pescador, and P. Corcoran, “Multi-phase obfuscation of fault secured DSP designs with enhanced security feature,” *IEEE Transactions on Consumer Electronics*, vol. 64, no. 3, pp. 356–364, 2018, doi:10.1109/TCE.2018.2852264.

R. Yasaei, L. Chen, S.-Y. Yu, and M. Abdullah Al Faruque, “Hardware Trojan detection using graph neural networks,” *IEEE Transactions on Computer-Aided Design of Integrated Circuits and Systems*, vol. 44, no. 1, pp. 25–38, Jan. 2025, doi:10.1109/TCAD.2022.3178355.

X. Cui, K. Ma, L. Shi, and K. Wu, “High-level synthesis for run-time hardware Trojan detection and recovery,” *2014 51st ACM/EDAC/IEEE Design Automation Conference (DAC)*, San Francisco, CA, USA, 2014, pp. 1–6.

S. Rajmohan, N. Ramasubramanian, and N. Naganathan, “Hybrid evolutionary design space exploration algorithm with defence against third party IP vulnerabilities,” *IEEE Transactions on Computer-Aided Design of Integrated Circuits and Systems*, vol. 39, no. 10, pp. 2602–2614, 2020, doi:10.1109/TCAD.2019.2960340.

A. Sengupta, A. Anshul, V. Chourasia, and N. Kumar, “M-HLS: malevolent high-level synthesis for watermarked hardware IPs,” *IEEE Embedded Systems Letters*, vol. 16, no. 4, pp. 497–500, 2024, doi:10.1109/LES.2024.3416422.

Silvaco’s Open Cell Library [Online], Available: https://si2.org/open-cell-and-free-pdk-libraries/, last accessed on December 2025.

A. Nahiyan, and M. Tehranipoor, “Code coverage analysis for IP trust verification,” In: Mishra, P., Bhunia, S., Tehranipoor, M (eds.), *Hardware IP Security and Trust*. Springer, Cham, 2017, doi:10.1007/978-3-319-49025-0_4.

M. Rostami, F. Koushanfar, and R. Karri, “A primer on hardware security: models, methods, and metrics,” *Proceedings IEEE*, vol. 102, no. 8, pp. 1283–1295, 2014, doi:10.1109/JPROC.2014.2335155.

A. Sengupta, and N. Bhui, “Time-bomb HLS Trojan for performance degradation payload,” *IEEE Design & Test*, vol. 42, no. 5, pp. 83–92, 2025a, doi:10.1109/MDAT.2025.3575356.

A. Sengupta, “Hardware security of CE devices [Hardware Matters],” *IEEE Consumer Electronics Magazine*, vol. 6, no. 1, pp. 130–133, 2017a, doi:10.1109/MCE.2016.2614552.

A. Sengupta, “Hardware vulnerabilities and their effects on CE devices: design for security against Trojans [Hardware Matters],” *IEEE Consumer Electronics Magazine*, vol. 6, no. 3, pp. 126–133, 2017b, doi:10.1109/MCE.2017.2684940.

A. Sengupta, and N. Bhui, "Securing hardware IP core for CE systems using fingerprint biometric integrated with microscopic fingerprint features and design encoded hash," *IEEE Transactions on Consumer Electronics*, vol. 71, no. 3, pp. 7983–7995, 2025b, doi:10.1109/TCE.2025.3593505.

A. Sengupta, N. Bhui, and A. Anshul, "High-Level Synthesis-Based Forensic Watermarking of Hardware IPs Using IP Vendor's DNA Signature," *IEEE Embedded Systems Letters*, vol. 18, no. 1, pp. 19–22, 2026, doi:10.1109/LES.2025.3566674.

M. Rathor, and A. Sengupta, "Revisiting Black-Hat HLS: a lightweight countermeasure to HLS-aided Trojan attack," *IEEE Embedded Systems Letters*, vol. 16, no. 2, pp. 170–173, 2024, doi:10.1109/LES.2023.3327793.

N. Bhui, and A. Sengupta, "Leveraging multi-layer crypto-logic cross-coupled with encoded code converter and conditional control flow for robust key-based HLS IP watermarking," *2025 IEEE 38th International System-on-Chip Conference (SOCC)*, Dubai, United Arab Emirates, 2025, pp. 1–6, doi:10.1109/SOCC66126.2025.11235366.

Y. Jin, and Y. Makris, "Hardware Trojan detection using path delay fingerprint," *2008 IEEE International Workshop on Hardware-Oriented Security and Trust*, Anaheim, CA, 2008, pp. 51–57, doi:10.1109/HST.2008.4559049.

A. Sengupta, V. Chourasia, and N. Bhui, "HLS Trojan classification and its detection techniques for CAD-based IP designs," *IT Professional*, vol. 27, no. 5, pp. 66–74, 2025c, doi:10.1109/MITP.2025.3584043.

A. Sengupta, A. Anshul, and R. Chaurasia, "Exploration of optimal functional trojan-resistant hardware intellectual property (IP) core designs during high level synthesis," *Microprocessors Microsystems*, vol. 103, 2023, Art. no. 104973, doi:10.1016/j.micpro.2023.104973.

A. Sengupta, and N. Bhui, "HLS watermarking of IP designs using scheduling driven key-based parallel switching framework integrated with multimodal crypto logic," *IEEE Embedded Systems Letters*, 2025c, doi:10.1109/LES.2025.3608556.

A. Anshul, and A. Sengupta, "A survey of high level synthesis based hardware security approaches for reusable IP cores [Feature]," *IEEE Circuits and Systems Magazine*, vol. 23, no. 4, pp. 44–62, Fourth quarter 2023, doi:10.1109/MCAS.2023.3325607.

A. Sengupta, V. Chourasia, N. Bhui, and A. Anshul, "Watermarking of transient fault-detectable IP designs using multivariate HLS scheduling based multimodal security," *IET Computers & Digital Techniques*, vol. 2025, Article ID 5926846, 2025d, doi:10.1049/cdt2/5926846.

N. Veeranna, and B. C. Schafer, "Hardware Trojan detection in behavioral intellectual properties (IP's) using property checking techniques," *IEEE Transactions on Emerging Topics in Computing*, vol. 5, no. 4, pp. 576–585, 2017, doi:10.1109/TETC.2016.2585046.

A. Sengupta, and S. Bhadauria, "Exploring low cost optimal watermark for reusable IP cores during high level synthesis," *IEEE Access*, vol. 4, pp. 2198–2215, 2016, doi:10.1109/ACCESS.2016.2552058.

A. Sengupta, and S. Bhadauria. "Untrusted third party digital IP cores: power-delay trade-off driven exploration of hardware Trojan secured datapath during high level synthesis," *Proceedings of the 25th edition on Great Lakes Symposium on VLSI*, pp. 167–172, 2015, doi:10.1145/2742060.2742061.

A. Sengupta, and M. Rathor, "IP core steganography for protecting DSP kernels used in CE systems," *IEEE Transactions on Consumer Electronics*, vol. 65, no. 4, pp. 506–515, 2019, doi:10.1109/TCE.2019.2944882.

A. Sengupta, and M. Rathor, "Securing hardware accelerators for CE systems using biometric fingerprinting," *IEEE Transactions on Very Large Scale Integration (VLSI) Systems*, vol. 28, no. 9, pp. 1979–1992, 2020, doi:10.1109/TVLSI.2020.2999514.

Part III

Detection and countermeasures

Chapter 6

Machine learning-based detection of HLS Trojans

Anirban Sengupta[1] *and Aditya Anshul*[1]

6.1 Introduction

Machine learning (ML) is a branch of artificial intelligence that empowers systems to learn patterns from data and make predictions or decisions without being explicitly programmed. Unlike traditional rule-based systems, ML models adapt through experience, improving their performance as more data becomes available. Broadly, ML techniques can be classified into supervised learning, unsupervised learning, and reinforcement learning (Hastie *et al.*, 2009). In supervised learning, models are trained on labeled data to map inputs to outputs, with algorithms such as linear regression, decision trees, support vector machines (SVMs), and neural networks being widely used. Linear regression is often applied in financial forecasting or demand prediction, while decision trees provide interpretable models for domains such as healthcare diagnostics and fraud detection. SVMs are effective in high-dimensional spaces, for example, in image classification or malware detection (Bishop, 2006). In unsupervised learning, where data lacks labels, techniques like clustering (e.g., k-means) or dimensionality reduction (e.g., principal component analysis) help uncover hidden structures in large datasets. Reinforcement learning, on the other hand, focuses on sequential decision-making, where an agent learns to maximize rewards through interaction with an environment, as seen in robotics or game-playing agents (Sutton and Barto, 2018).

In the field of hardware security, machine learning has recently gained traction as an effective tool for hardware Trojan detection. A hardware Trojan is a malicious modification of an integrated circuit (IC) that may remain dormant under normal conditions but activates under rare triggers to degrade performance, leak information, or disrupt functionality (Rostami *et al.*, 2014; Sengupta *et al.*, 2025), etc. Traditional state-of-the-art approaches for Trojan detection, such as functional validation, equivalence checking, and path delay fingerprinting, often suffer from low coverage because Trojans can remain stealthy and are triggered only under rare events (Anshul and Sengupta, 2025; Jin and Markis, 2008). Machine learning addresses this challenge

[1]Department of Computer Science and Engineering, Indian Institute of Technology Indore, India

by analyzing large volumes of design or features to identify subtle patterns indicative of Trojan insertion. By learning from both "safe/secure" and "infected" hardware designs, ML-based techniques significantly improve detection coverage and robustness. The ability to generalize across diverse Trojan types makes ML particularly attractive compared to conventional rule-based methods.

The importance of ML-driven Trojan detection becomes even more critical in the context of high-level synthesis (HLS) framework-generated intellectual property (IP). HLS enables the conversion of high-level algorithmic descriptions (e.g., C/C++) into register transfer level (RTL) designs, streamlining hardware design (Pearce *et al.*, 2014; Anshul and Sengupta, 2023a; Rathor *et al.*, 2014; Anshul and Sengupta, 2023b). However, this abstraction also introduces vulnerabilities, *viz.*, compromised HLS frameworks can contain secretly injected Trojans at multiple stages, such as the scheduling, resource allocation, and interconnect design stages. Six known types of HLS Trojans (documented in the state-of-the-art) are performance degradation (PD-HT) (Sengupta *et al.*, 2024), denial-of-service (DoS-HT) (Sengupta *et al.*, 2024), battery exhaustion (BE-HT) (Pilato *et al.*, 2019), downgrade attack (DA-HT) (Pilato *et al.*, 2019), functional (F-HT) (Sengupta *et al.*, 2017), and time-bomb (TB-HT) (Sengupta and Bhui, 2025), each with unique triggers and payloads (Anshul and Sengupta, 2025). Many of these Trojans evade state-of-the-art detection because they either do not alter functionality or activate only under rare conditions. State-of-the-art ML-based solutions (Anshul and Sengupta, 2025), however, can extract meaningful RTL-level features and classify designs as Trojan-free or infected with high accuracy, thereby closing gaps left by earlier methods.

This chapter focuses on decision tree-based machine learning algorithms for detecting HLS Trojans that have been recently presented by the authors in Anshul and Sengupta (2025). The motivation for choosing decision trees stems from their lightweight nature, interpretability, and computational efficiency (Anshul and Sengupta, 2025). Unlike SVMs, which require kernel tricks for nonlinear problems, or neural networks, which demand substantial computational resources and hyper-parameter tuning, decision trees operate efficiently with minimal preprocessing. They handle categorical and numerical data gracefully, manage collinearity better than many alternatives, and do not require normalization or scaling. These advantages make them particularly suitable for resource-constrained scenarios where lightweight yet effective models are essential. In the context of HLS-Trojan detection, decision trees excel at capturing nonlinear interactions among features such as comparator chains, multiplexer interconnections, counters, and tri-state buffers—all of which are potential indicators of Trojan logic. Recent research has demonstrated the effectiveness of the decision tree classifier in detecting all six major HLS Trojan types with high accuracy and zero false positives (Anshul and Sengupta, 2025).

6.2 Summary of the chapter

Hardware Trojans introduced in IP designs generated through HLS pose critical threats to hardware security. As discussed above, although HLS tools accelerate the transformation of high-level specifications into RTL implementations, they can also open avenues (attack surface) for adversaries to insert hidden malicious logic. The risks become even more pronounced when compromised HLS frameworks are employed in the design of systems used in national security, as they may implant undetectable vulnerabilities throughout the design process. This chapter discusses an HLS-Trojan detection approach that leverages a decision tree-based machine learning model to identify Trojans in HLS-generated RTL designs (Anshul and Sengupta, 2025). By extracting and analyzing distinctive RTL features, the framework achieves accurate classification of infected and non-infected designs. The analytical validation (results) demonstrates a high detection rate, including zero false positives for the Trojan categories studied, thereby offering a robust and efficient method for detecting Trojans in the IP designs.

6.3 Salient features of the chapter

- This chapter discusses an HLS-Trojan detection framework capable of identifying multiple types of covertly inserted hardware Trojans in HLS-based IP designs, including performance degradation, denial-of-service, battery exhaustion, downgrade attack, functional, and time-bomb Trojans (Anshul and Sengupta, 2025).
- A dedicated feature extraction block (Anshul and Sengupta, 2025) has been discussed, which automatically extracts critical RTL-level characteristics/features from hardware IP RTL designs (in the form of VHDL), which serve as essential inputs for effective Trojan detection.
- A decision tree-machine learning model-based HLS-Trojan detection framework has been discussed that provides an efficient and interpretable method for distinguishing between Trojan-infected and clean IP designs.
- Analytical validation highlights that the discussed ML-based HLS-Trojan detection approach (Anshul and Sengupta, 2025) depicts high detection accuracy across all six considered Trojan types, thereby ensuring reliable and trustworthy hardware security.

6.4 Discussion on the threat model

The threat model in this chapter focuses on the vulnerabilities introduced by untrustworthy (third-party) HLS frameworks/tools during hardware IP design. HLS tools automate the generation of RTL circuits from high-level code, where each stage of the HLS process is considered untrustworthy and a potential vulnerability point for an adversary to inject backdoor malicious logic (Anshul and Sengupta, 2025; Sengupta *et al.*, 2024; Pilato *et al.*, 2019). Since these HLS tools are developed by a third-party vendor, an adversary present inside the HLS tool vendor house can exploit internal HLS stages such as scheduling, resource allocation, controller/datapath synthesis, or multiplexer (Mux)-interconnect design to implant hardware Trojans (that remain hidden until rare conditions activate them). These Trojans are dangerous because they often bypass conventional detection methods, such as functional testing, equivalence checking, or side-channel analysis, by appearing dormant under normal execution. For example, while many HLS tools provide RTL functional validation and some include C-to-RTL equivalence checking, these methods are ineffective against most HLS Trojans. This is because such Trojans either do not impact functionality at all or activate only under rare trigger conditions, leaving them dormant during normal execution. Moreover, C-to-RTL equivalence checking requires access to Trojan-infected C-code, which is not always available, since Trojans may be inserted at different stages of the HLS flow (Anshul and Sengupta, 2025; Sengupta *et al.*, 2024; Pilato *et al.*, 2019). Further, most of the HLS tools do not offer provisions for C-to-RTL equivalence checking.

Furthermore, once triggered, they can lead to serious consequences, including performance degradation, denial-of-service, excessive power consumption, weakened cryptographic-based protection, or functional/data corruption of the hardware design. The problem is further exacerbated when compromised HLS tools, possibly distributed across international supply chains, are used in the IP designs of system-on-chip (SoC). For example, compromised HLS frameworks may serve as part of a national-level attack arsenal, where untrusted tools/generated IPs are distributed across multiple countries. These IPs, as part of SoC designs, may contain covert backdoor Trojans, without the knowledge of the original IP designer/vendor. Moreover, the RTL IP designs produced by HLS are significantly larger and more complex than their high-level counterparts, making manual inspection impractical (Anshul and Sengupta, 2025; Sengupta *et al.*, 2024; Pilato *et al.*, 2019). Together, these factors establish HLS Trojans as a critical security threat, motivating the need for automated and intelligent detection approaches such as the ML-Trojan detection framework, as discussed in this chapter.

6.5 Classification of HLS Trojans and their detection using ML

Figure 6.1 depicts the overview of the HLS-Trojan detection methodology using a machine learning (ML) framework (Anshul and Sengupta, 2025). The primary input consists of RTL descriptions of hardware IPs (VHDL code)—Trojan-infected HLS-based RTL/Non-infected HLS-based RTL. The final output is the detection of a hardware Trojan (potentially present in an infected IP RTL). The major blocks of ML-Trojan detection methodology are: (a) feature extraction block—responsible for extracting important unique features from the HLS-generated IP RTL, (b) database formation for the ML framework—used for training the ML model, and (c) Trojan detection block based on a decision tree-based ML classifier. The details of each block are discussed below.

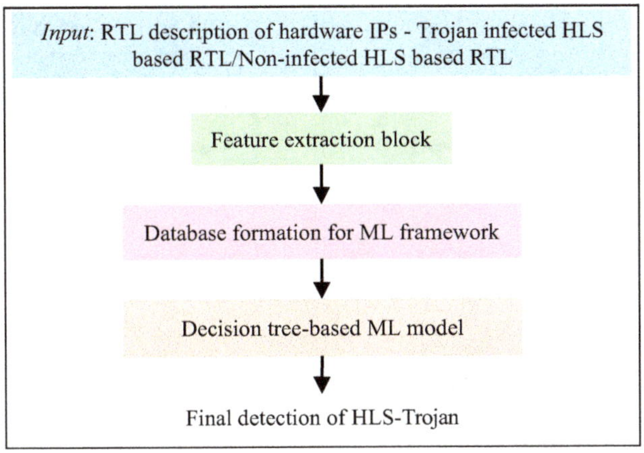

Figure 6.1 Overview of the ML-based HLS-Trojan detection technique

Classification of HLS Trojans: Hardware Trojans can take various forms, each with specific malicious objectives. The six different classes of HLS Trojans have already been mentioned in the introduction section of this chapter. As per the name, performance degradation hardware Trojans (PD-HTs) are designed to impair system performance (performance degradation) when triggered under a predefined condition. Next, denial-of-service hardware Trojans (DoS-HTs) disrupt operation, causing an open circuit/high-impedance state or indeterminate condition. These Trojans rely on comparator logic to activate under a specific, predefined, rare condition, making them difficult to detect. Subsequently, battery exhaustion attack/Trojan (BE-HT) modifies the hardware design datapath to increase power consumption, accelerating battery discharge (power degradation) while evading power side-channel analysis, often by utilizing idle units with high dynamic power usage (such as multipliers). Next, the downgrade attack/Trojan (DA-HT) weakens the overall security of crypto-cores by reducing cryptographic algorithm rounds. These Trojans are triggered using a counter-based trigger (counter) that activates the Trojan after a predefined number of executions and input sequence detectors, respectively. Next, functional Trojans (F-HT) alter output values of micro 3rd party IP (3PIP) as part of the HLS module library on rare condition triggering, activated via mechanisms such as FSM counter/external signal. Furthermore, time-bomb hardware Trojans (TB-HT) are also designed to impair system performance (performance degradation) when triggered under a predefined condition using integrated counter and comparator logic. These diverse Trojan types pose significant challenges to detecting malice in an HLS-generated IP core. Figure 6.2. highlights an overview of HLS-Trojan types along with their insertion stages, corresponding triggering mechanisms, and payloads.

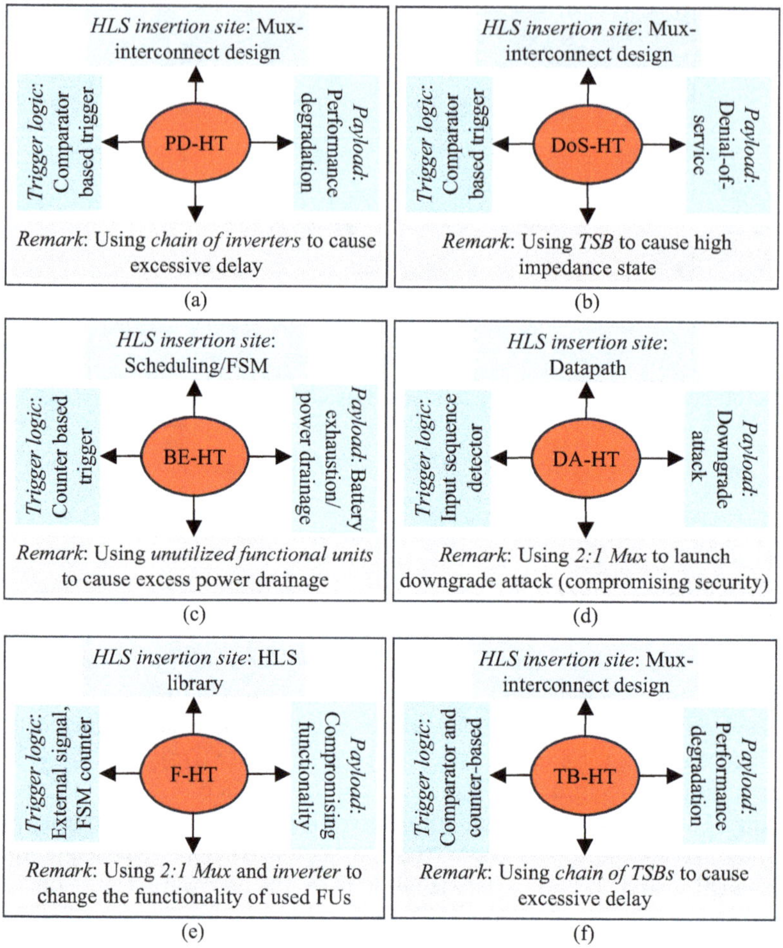

Figure 6.2 HLS-Trojan types along with their insertion stages, corresponding triggering mechanism, and payload

6.6 Feature extraction process corresponding to different HLS Trojans and the ML detection methodology

The details of the ML-based HLS-Trojan detection approach (Anshul and Sengupta, 2025) are illustrated in Figure 6.3. As depicted in Figure 6.3, the ML-Trojan detection framework begins with the RTL descriptions of hardware IPs (both Trojan-free (clean) and Trojan-infected) expressed in VHDL code. These RTL IP designs are first processed through a custom feature extraction module that identifies structural and behavioral markers indicative of hardware Trojans. The extracted features are then categorized into two sets: those corresponding to

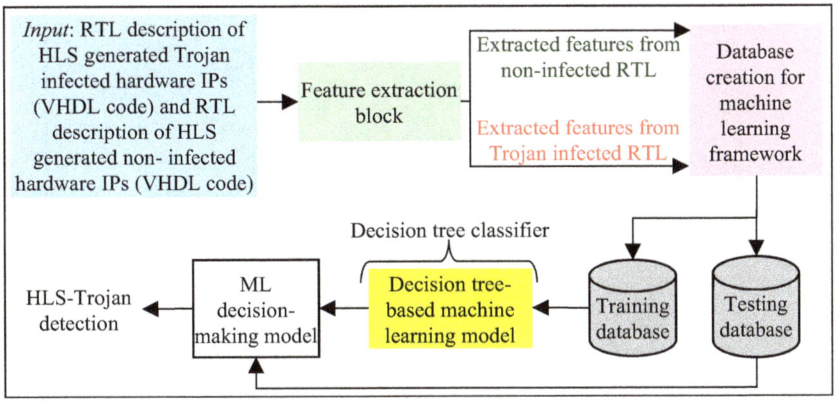

Figure 6.3 Details of the ML-based HLS-Trojan detection approach

infected designs and those from non-infected ones. These feature vectors form the dataset for training and validating/testing a machine learning classifier/model. As discussed in the introduction section of this chapter, a decision tree model has been chosen for this task owing to its interpretability, low computational overhead, and ability to capture nonlinear feature interactions without extensive preprocessing. Once trained, the classifier distinguishes between Trojan-free RTLs and those carrying malicious modifications (Trojans).

Feature extraction module: The pseudocode corresponding to the custom feature extraction module (Anshul and Sengupta, 2025) is shown in Figure 6.4. Each type of HLS Trojan exhibits a distinct set of structural cues within the RTL, which the feature extraction process systematically identifies. For the PD-HT, the relevant signatures include the presence of long chains of inverters, comparators, and multiplexer outputs feeding into other multiplexers, all of which indicate intentional propagation delay insertion and trigger logic, respectively. Similarly, the TB-HT is detected through a counter and a comparator combined with tri-state buffer chains and a nested multiplexer path, reflecting its activation and performance impact. Subsequently, DoS-HT is recognized by comparator-triggered activation logic along with tri-state buffers that drive modules into high-impedance states, effectively stalling the output (normal operation). Next, the BE-HT introduces counters to activate idle functional units, increasing dynamic power usage; hence, counters and redundant hardware activations are primary features here. In the case of the DA-HT, input sequence detector and comparator-driven trigger logic alter cryptographic routines by reducing the number of valid iterations, with evidence observed in comparator and multiplexer modifications. Finally, the F-HT exploits an extra input port in the IP design (which may be considered a suspicious input port/signal in the design) to hijack control logic or alter computation outcomes. The point-wise crucial features are mentioned below for each HLS Trojan:

Input: RTL description of HLS generated Trojan infected hardware IPs (VHDL code) and RTL description of HLS generated non-infected hardware IPs (VHDL code)
Output: Feature set corresponding to input RTL

⇨ Input and output

```
X ← RTL file (VHDL)
Ψ ← Port map information present in X
    FOR each line (i) in X
        IF (mux_2_to_1 ∃ i)
            2*1 Mux = 2*1 Mux + 1
        IF (mux_4_to_1 ∃ i)
            4*1 Mux = 4*1 Mux + 1
        IF (mux_8_to_1 ∃ i)
            8*1 Mux = 8*1 Mux + 1
        IF (mux_16_to_1 ∃ i)
            16*1 Mux = 16*1 Mux + 1
        IF (mux_32_to_1 ∃ i)
            32*1 Mux = 32*1 Mux + 1
        IF (demux_2_to_1 ∃ i)
            2*1 DeMux = 2*1 DeMux + 1
        IF (demux_4_to_1 ∃ i)
            4*1 DeMux = 4*1DeMux + 1
        IF (demux_8_to_1 ∃ i)
            8*1 DeMux = 8*1 DeMux + 1
        IF (demux_16_to_1 ∃ i)
            16*1 DeMux = 16*1 DeMux + 1
        IF (demux_32_to_1 ∃ i)
            32*1 DeMux = 32*1 DeMux + 1
```

⇨ Mux-interconnect information and potential Trojan logic in RTL IP

```
        IF (latch ∃ i)
            Latch = Latch + 1
```

⇨ Storage unit

```
        IF (Adder ∃ i)
            Adder = Adder + 1
        IF (Subtractor ∃ i)
            Subtractor = Subtractor + 1
        IF (Multiplier ∃ i)
            Multiplier = Multiplier + 1
```

⇨ Functional unit information

```
        IF (Comparator ∃ i)
            comparator = 1
        IF (up_counter ∃ i)
            Up-counter = 1
        IF (tri_state_buffer ∃ i)
            Tri-state buffer = 1
        END IF
    END FOR
IF (Output of TSB → input of TSB in Ψ)
        TSB_Chain = total TSB in the TSB chain
IF (Output of Mux → input of another Mux in Ψ)
        Mux output acting as input for another Mux = 1
IF (Output of NOT gate → input of another NOT gate in Ψ)
        Inverter Chain = total inverters in the inverter chain
IF (Third input port of a component (functional unit) ∃ Ψ)
        Third_FU_Input = 1
```

⇨ Potential Trojan logic in RTL IP

Figure 6.4 Pseudocode corresponding to the feature extraction algorithm

- ***Performance Degradation Trojan (PD-HT):***
  - Comparator and "Multiplexer output driving another multiplexer" used as trigger logic.
  - Chains of inverters inserted to increase propagation delay.

- ***Time-Bomb Trojan (TB-HT):***
  - Counter-based trigger combined with comparator and nested multiplexer connections used as part of activation logic.
  - Chains of tri-state buffers (TSBs) causing performance degradation.
- ***Denial-of-Service Trojan (DoS-HT):***
  - Nested multiplexer connections and a comparator-driven triggering condition.
  - Presence of tri-state buffers that force high-impedance states.
- ***Battery Exhaustion Trojan (BE-HT):***
  - Up-counter serving as an activation mechanism after a threshold number of executions.
  - Redundant functional unit utilization to artificially increase power drainage.
- ***Downgrade Attack Trojan (DA-HT):***
  - Input sequence detector and comparator-based triggering mechanism.
  - Enforcing reduced iterations for compromising cryptographic round counts.
  - Evidence of manipulation using a multiplexer to achieve a downgrade.
- ***Functional Trojan (F-HT):***
  - FSM counters or external control signals used as activation triggers.
  - Suspicious extra input port in the IP design, which may be exploited for malicious control.
  - Alterations using an inverter and multiplexer to affect computation outputs.

HLS-Trojans dataset and ML classification: Authors Anshul and Sengupta (2025) have considered (extracted) twenty-one RTL features across different Trojan-infected and non-infected designs. Each RTL instance is labeled according to its class: Trojan-free (0), PD-HT (1), DoS-HT (2), BE-HT (3), DA-HT (4), F-HT (5), and TB-HT (6). These labeled feature sets constitute the database used to train and evaluate the decision tree model. Through this process, the framework ensures effective detection of all six Trojans, thereby providing a comprehensive ML-based defense against HLS-inserted hardware Trojans. Figure 6.5 depicts the snippet of the database (containing 21 features). The database contains instances corresponding to CNN convolutional layer IP, FIR Filter IP, DCT IP, IIR Filter IP, GLRT IP, MESA IP, and Sharpening Filter IP benchmarks. The complete database, along with the details corresponding to the used benchmarks, is available at the GitHub Repository (2025). Next, an infected VHDL code corresponding to the PD-HT infected FIR IP design is shown in Figure 6.6. Subsequently, Figure 6.7 depicts the RTL schematic corresponding to the PD-HT infected FIR IP design (shown in Figure 6.6). Finally, the complete Python code corresponding to the feature extraction algorithm (shown in Figure 6.4) is illustrated in the appendix section (at the end of this chapter). Moreover, the Trojan-infected VHDL code, PD-HT-infected FIR RTL schematic, and the Python code corresponding to the custom feature extraction module are available at the GitHub Repository (2025).

1		4*1 Mux	8*1 Mux	16*1 Mux	32*1 Mux	1*2 DeMux	1*4 DeMux	1*8 DeMux	1*16 DeMux	1*32 DeMux	Latch	Adder	Subtractor	Multiplier	Comparator	2*1 Mux	TSB chain	Inverter Chain	Up-counter	Tri-state buffer	acting as input for :	Third_FU_Input	Label
2	Covolutional layer-NI	0	4	2	0	8	0	3	1	0	6	1	0	1	0	8	0	0	0	0	0	0	0
3	Covolutional layer-I1	0	4	2	0	8	0	3	1	0	6	1	0	1	1	11	0	20	0	0	1	0	1
4	Covolutional layer-I2	0	4	2	0	8	0	3	1	0	6	1	0	1	1	11	0	0	0	1	1	0	2
5	Covolutional layer-I3	0	4	2	0	8	0	3	1	0	6	1	0	1	0	8	0	0	1	0	0	0	3
6	Covolutional layer-I4	0	4	2	0	8	0	3	1	0	6	1	0	1	1	9	0	0	0	0	0	0	4
7	Covolutional layer-I5	0	4	2	0	8	0	3	1	0	6	1	0	1	0	9	0	0	0	0	0	1	5
8	Covolutional layer-I6	0	4	2	0	8	0	3	1	0	6	1	0	1	1	11	48	0	1	0	1	0	6
9	FIR-NI	12	2	2	0	0	10	1	1	0	12	2	0	2	0	0	0	0	0	0	0	0	0
10	FIR-I1	12	2	2	0	0	10	1	1	0	12	2	0	2	1	3	0	20	0	0	1	0	1
11	FIR-I2	12	2	2	0	0	10	1	1	0	12	2	0	2	1	3	0	0	0	1	1	0	2
12	FIR-I3	12	2	2	0	0	10	1	1	0	12	2	0	2	0	0	0	0	1	0	0	0	3
13	FIR-I4	12	2	2	0	0	10	1	1	0	12	2	0	2	1	1	0	0	0	0	0	0	4
14	FIR-I5	12	2	2	0	0	10	1	1	0	12	2	0	2	0	1	0	0	0	0	0	1	5
15	FIR-I6	12	2	2	0	0	10	1	1	0	12	2	0	2	1	3	48	0	1	0	1	0	6
16	MESA-NI	3	1	0	0	7	3	1	0	0	21	3	0	4	0	14	0	0	0	0	0	0	0
17	MESA-I1	3	1	0	0	7	3	1	0	0	21	3	0	4	1	17	0	20	0	0	1	0	1
18	MESA-I2	3	1	0	0	7	3	1	0	0	21	3	0	4	1	17	0	0	0	1	1	0	2
19	MESA-I3	3	1	0	0	7	3	1	0	0	21	3	0	4	0	14	0	0	1	0	0	0	3
20	MESA-I4	3	1	0	0	7	3	1	0	0	21	3	0	4	1	15	0	0	0	0	0	0	4
21	MESA-I5	3	1	0	0	7	3	1	0	0	21	3	0	4	0	15	0	0	0	0	0	1	5
22	MESA-I6	3	1	0	0	7	3	1	0	0	21	3	0	4	1	17	48	0	1	0	1	0	6
23	GLRT-NI	6	0	0	0	2	5	0	0	0	9	2	0	1	0	4	0	0	0	0	0	0	0
24	GLRT-I1	6	0	0	0	2	5	0	0	0	9	2	0	1	1	7	0	20	0	0	1	0	1
25	GLRT-I2	6	0	0	0	2	5	0	0	0	9	2	0	1	1	7	0	0	0	1	1	0	2
26	GLRT-I3	6	0	0	0	2	5	0	0	0	9	2	0	1	0	4	0	0	1	0	0	0	3
27	GLRT-I4	6	0	0	0	2	5	0	0	0	9	2	0	1	1	5	0	0	0	0	0	0	4

Figure 6.5 *Snippet of the HLS-Trojan dataset (available at [16]). Note: Here, rows indicate the instances corresponding to the different benchmarks, and columns indicate 21 different features along with the label. In the HLS-Trojan database (available at [16]), row #2–row #50 indicates the seven unique instances (six Trojan-infected and one non-Trojan-infected) corresponding to each benchmark. Additionally, row #51–row # 78 indicates the variations of instance #2 (PD-HT infected IP designs).*

```
library IEEE;
use IEEE.STD_LOGIC_1164.ALL;
use IEEE.STD_LOGIC_ARITH.ALL;
use IEEE.STD_LOGIC_UNSIGNED.ALL;

entity FIRIP is
Port (IN1,IN2,IN3,IN4,IN5,IN6,IN7,IN8,IN10,
      S1,S2,S3,S4,S5,S6,S7,S8,S9,S10,S11,S12,S13,S14,
              S22,S23: in std_logic_vector(15 downto 0);

      S4_1,S4_2,S4_3:in std_logic_vector(3 downto 0);

              SEL_A2,DSEL_A2:in std_logic_vector(2 downto 0);

              IN9,IN11,SEL_M1,DSEL_M1,SEL_M2,DSEL_M2,SEL_DB,DSEL_DB,SEL_LB,DSEL_LB,SEL_G,DSEL_G,SEL_Y,DSEL_Y,SEL_R,
              DSEL_R,SEL_O,DSEL_O,SEL_B,DSEL_B:in std_logic_vector(1 downto 0);

      X1, X2, X3, X4, X5, X6, X7, X8, X9,X10,X11,X12,X13,X14,X15,X16,X17: in std_logic;

              FO,F1,F2,S15,S16,S17,S18,S19,S20,S21: out std_logic_vector(15 downto 0));
end FIRIP;
```

(a) VHDL library and port declaration code snippet

```
architecture structural of FIRIP is

component comparator is
port(A1,B1: in std_logic_vector(15 downto 0);
 A_equal_B: out std_logic);
 end component;

component mux_2to1 is
port(A2,B2: in std_logic_vector(1 downto 0);
          MS2: in std_logic;
          MZ2: out std_logic_vector(1 downto 0));
end component;

component mux2to1 is
port(A21,B21: in std_logic_vector(15 downto 0);
           MS21: in std_logic;
           MZ21: out std_logic_vector(15 downto 0));
end component;

component mux_4to1 is
port(A4,B4,C4,D4:in std_logic_vector(15 downto 0);
        MS4: in std_logic_vector(1 downto 0);
        MZ4: out std_logic_vector(15 downto 0));
end component;
```

(b) VHDL component declaration code snippet

```
component mux_8to1 is
port(F0,F1,F2,F3,F4,F5,F6,F7: in std_logic_vector(15 downto 0);
     MS8: in std_logic_vector(2 downto 0);
     MZ8: out std_logic_vector(15 downto 0));
end component;

component mux_16to1 is
port(A16,B16,C16,D16,E16,F16,G16,H16,I16,J16,K16,L16,M16,N16,O16,
         P16: in std_logic_vector(15 downto 0);
         MS16: in std_logic_vector(3 downto 0);
         MZ16: out std_logic_vector(15 downto 0));
end component;

component adder is
port(aA: in std_logic_vector(15 downto 0);
     aB: in std_logic_vector(15 downto 0);
         aZ: out std_logic_vector(15 downto 0));
end component;

component mul is
port(mA: in std_logic_vector(15 downto 0);
     mB: in std_logic_vector(15 downto 0);
     mZ: out std_logic_vector(15 downto 0));
end component;
```

(c) VHDL component declaration code snippet

```
component demux_1to2 is
port(Z2: in std_logic_vector(15 downto 0);
         DS2: in std_logic;
  P2,Q2:out std_logic_vector(15 downto 0));
end component;

component demux_1to4 is
port(P4,Q4,R4,T4:out std_logic_vector(15 downto 0);
             DS4: in std_logic_vector(1 downto 0);
             DZ4: in std_logic_vector(15 downto 0));
end component;

component demux_1to8 is
port(DZ8: in std_logic_vector(15 downto 0);
          DS8: in std_logic_vector(2 downto 0);
     P0,P1,P2,P3,P4,P5,P6,P7: out std_logic_vector(15 downto 0));
end component;

component demux_1to16 is
Port (DZ16: in std_logic_vector(15 downto 0);
               DS16: in std_logic_vector(3 downto 0);
      Y0,Y1,Y2,Y3,Y4,Y5,Y6,Y7,Y8,Y9,Y10,Y11,Y12,Y13,Y14,
        Y15: out std_logic_vector(15 downto 0));
end component;
```

(d) VHDL component declaration code snippet

```
component latch1 is
port(L1: in std_logic_vector(15 downto 0);
   LEN1: in std_logic;
    LZ1: out std_logic_vector(15 downto 0));

end component;
component latch2 is
port(L2: in std_logic_vector(15 downto 0);
   LEN2: in std_logic;
     W1: out std_logic_vector(15 downto 0);
          W2: out std_logic_vector(15 downto 0));
end component;

component reg16 is
port(rD: in std_logic_vector(15 downto 0);
          ren: in std_logic;
            rZ: out std_logic_vector(15 downto 0));
end component;

component NOT2 is
port (X: in std_logic_vector(15 downto 0);
       Y: out std_logic_vector(15 downto 0));
end component;
```

(e) VHDL component declaration code snippet

```
SIGNAL S0: std_logic;
SIGNAL SEL_P,DSEL_P: std_logic_vector(1 downto 0);

SIGNAL C,S15_1,S15_2,S15_3,S15_4,S15_5,S15_6,S15_7,S15_8,S15_9,S15_10,
       S15_11,S15_12,S15_13,S15_14,S15_15,S15_16,S15_17,S15_18,S15_19,S15_20,
       S15_21,S15_22,S15_23,S15_24,S15_25,S15_26,S15_27,S15_28,S15_29,S15_30,
       S15_31,S15_32,S15_33,S15_34,S15_35,S15_36,S15_37,S15_38,S15_39,S15_40,
       S15_41,S15_42,S15_43,S15_44,S15_45,S15_46,S15_47,S15_48,S15_49,S15_50,
       S15_51,S15_52,S15_53,S15_54,S15_55,S15_56,S15_57,S15_58,S15_59,S15_60,
       S15_61,S15_62,S15_63,S15_64,S15_65,S15_66,S15_67,S15_68,S15_69,S15_70,
              S15_71,S15_72,S15_73,S15_74,S15_75,S15_76,S15_77,S15_78,S15_79,S15_80,
              S15_81,S15_83,S15_84,S15_85,S15_86,S15_87,S15_88,S15_89,S15_90,
              S15_91,S15_92,S15_93,S15_94,S15_95,S15_96,S15_97: std_logic_vector(15 downto 0);
```

(f) VHDL intermediate signal declaration code snippet

Figure 6.6 (a)–(j). Infected VHDL code corresponding to the PD-HT infected FIR IP design

```
begin
--------------------------TROJAN BLOCK--------------REG P MUX 4X1----------------
U1:mux_4to1 PORT MAP(A4=>IN1,B4=>S15_1,C4=>S15_2,D4=>S15_6,MS4=>SEL_P,MZ4=>S15_7);

U2:NOT2 PORT MAP(X=>IN1,Y=>S15_3);

U3:NOT2 PORT MAP(X=>S15_3,Y=>S15_4);

U4:NOT2 PORT MAP(X=>S15_4,Y=>S15_5);

U5:NOT2 PORT MAP(X=>S15_5,Y=>S15_6);

U6:mux_2to1 PORT MAP(A2=>IN9,B2=>not(IN9),MS2=>S0,MZ2=>SEL_P);
U7:reg16 PORT MAP(rD=>IN10,ren=>X9,rZ=>C);
U8:comparator PORT MAP(A1=>S15_54,B1=>C,A_equal_B=>S0);
U9:reg16 PORT MAP(rD=>S15_7,ren=>X1,rZ=>S15_8);
-----------------TROJAN BLOCK---------------REG P DEMUX 1X4----------------------
U10:demux_1to4 PORT MAP(P4=>S15_9,Q4=>S15_10,R4=>S15_11,T4=>S15_12,DS4=>DSEL_P,DZ4=>S15_8);
U11:mux_2to1 PORT MAP(A2=>IN11,B2=>not(IN11),MS2=>S0,MZ2=>DSEL_P);
U12:mux2to1 PORT MAP(A21=>S15_9,B21=>S15_12,MS21=>S0,MZ21=>S15_13);
```

(g) VHDL Trojan block port map code snippet

```
----------------------------REG DB-----------------------------------------------
U13:mux_4to1 PORT MAP(A4=>IN2,B4=>S15_14,C4=>S15_15,D4=>S15_16,MS4=>SEL_DB,MZ4=>S15_26);
U14:reg16 PORT MAP(rD=>S15_26, ren=>X2, rZ=>S15_30);
U15:demux_1to4 PORT MAP(P4=>S15_34,Q4=>S15_35,R4=>S15_36,T4=>S15_37,DS4=>DSEL_DB,DZ4=>S15_30);
----------------------------REG LB-----------------------------------------------
U16:mux_4to1 PORT MAP(A4=>IN3,B4=>S15_17,C4=>S15_18,D4=>S15_19,MS4=>SEL_LB,MZ4=>S15_27);
U17:reg16 PORT MAP(rD=>S15_27, ren=>X3, rZ=>S15_31);
U18:demux_1to4 PORT MAP(P4=>S15_38,Q4=>S15_39,R4=>S15_40,T4=>S15_41,DS4=>DSEL_LB,DZ4=>S15_31);
-----------------------------REG G-----------------------------------------------
U19:mux_4to1 PORT MAP(A4=>IN4,B4=>S15_20,C4=>S15_21,D4=>S15_22,MS4=>SEL_G,MZ4=>S15_28);
U20:reg16 PORT MAP(rD=>S15_28, ren=>X4, rZ=>S15_32);
U21:demux_1to4 PORT MAP(P4=>S15_42,Q4=>S15_43,R4=>S15_44,T4=>S15_45,DS4=>DSEL_G,DZ4=>S15_32);
-----------------------------REG Y-----------------------------------------------
U22:mux_4to1 PORT MAP(A4=>IN5,B4=>S15_23,C4=>S15_24,D4=>S15_25,MS4=>SEL_Y,MZ4=>S15_29);
U23:reg16 PORT MAP(rD=>S15_29, ren=>X5, rZ=>S15_33);
U24:demux_1to4 PORT MAP(P4=>S15_46,Q4=>S15_47,R4=>S15_48,T4=>S15_49,DS4=>DSEL_Y,DZ4=>S15_33);
-----------------------------A1 Block--------------------------------------------------
U25:mux_16to1 PORT MAP(A16=>S15_13,B16=>S15_38,C16=>S15_11,D16=>S15_44,E16=>S15_45,F16=>S15_41,G16=>S15_49,
   H16=>S15_74,I16=>S15_78,J16=>S1,K16=>S2,L16=>S3,M16=>S4,N16=>S5,O16=>S6,P16=>S7, MS16=>S4_1, MZ16=>S15_50);
U26:latch1 PORT MAP(L1=>S15_50,LEN1=>X17,LZ1=>S15_52);
U27:mux_16to1 PORT MAP(A16=>S15_13,B16=>S15_38,C16=>S15_36,D16=>S15_37,E16=>S15_40,F16=>S15_48,G16=>S15_73,
  H16=>S15_77,I16=>S15_81,J16=>S8,K16=>S9,L16=>S10,M16=>S11,N16=>S12,O16=>S13,P16=>S14,MS16=>S4_2, MZ16=>S15_51);
U28:latch1 PORT MAP(L1=>S15_51,LEN1=>X6,LZ1=>S15_53);
U29:adder PORT MAP(aA=>S15_52,aB=>S15_53,aZ=>S15_54);
U30:latch1 PORT MAP(L1=>S15_54,LEN1=>X7,LZ1=>S15_55);
U31:demux_1to16 PORT MAP(Y0=>S15_2,Y1=>S15_20,Y2=>S15_21,Y3=>S15_22,Y4=>S15_19,Y5=>S15_25,Y6=>S15_58,
  Y7=>S15_61,Y8=>S15_64,Y9=>S15,Y10=>S16,Y11=>S17,Y12=>S18,Y13=>S19,Y14=>S20,Y15=>S21,DS16=>S4_3,DZ16=>S15_55);
```

(h) VHDL register and functional unit port map code snippet

```
-------------------------------------REG GO--------------------------------------------------
U32:mux_4to1 PORT MAP(A4=>IN6,B4=>S15_56,C4=>S15_57,D4=>S15_58,MS4=>SEL_R ,MZ4=>S15_65);
U33:reg16 PORT MAP(rD=>S15_65, ren=>X8, rZ=>S15_68);
U34:demux_1to4 PORT MAP(P4=>S15_71,Q4=>S15_72,R4=>S15_73,T4=>S15_74,DS4=>DSEL_R,DZ4=>S15_68);
-------------------------------------REG O-------------------------------------------------
U35:mux_4to1 PORT MAP(A4=>IN7,B4=>S15_59,C4=>S15_60,D4=>S15_61,MS4=>SEL_O,MZ4=>S15_66);
U36:reg16 PORT MAP(rD=>S15_66, ren=>X10, rZ=>S15_69);
U37:demux_1to4 PORT MAP(P4=>S15_75,Q4=>S15_76,R4=>S15_77,T4=>S15_78,DS4=>DSEL_O,DZ4=>S15_69);
-------------------------------------REG B-------------------------------------------------
U38:mux_4to1 PORT MAP(A4=>IN8,B4=>S15_62,C4=>S15_63,D4=>S15_64,MS4=>SEL_B,MZ4=>S15_67);
U39:reg16 PORT MAP(rD=>S15_67, ren=>X11,rZ=>S15_70 );
U40:demux_1to4 PORT MAP(P4=>S15_79,Q4=>S15_80,R4=>S15_81,T4=>FO,DS4=>DSEL_B,DZ4=>S15_70);
-------------------------------------M1 Block----------------------------------------------
U41:mux_4to1 PORT MAP(A4=>S15_10,B4=>S15_43,C4=>S15_47,D4=>S15_76,MS4=>SEL_M1,MZ4=>S15_83);
U42:latch2 PORT MAP(L2=>S15_83,LEN2=>X12,W1=>S15_84,W2=>S15_85);

U43:mul PORT MAP(mA=>S15_84,mB=>S15_85,mZ=>S15_86);
U44:latch1 PORT MAP(L1=>S15_86,LEN1=>X13,LZ1=>S15_87);
U45:demux_1to4 PORT MAP(P4=>S15_1,Q4=>S15_16,R4=>S15_24,T4=>S15_60,DS4=>DSEL_M1,DZ4=>S15_87);
---------------------------M2 Block-------------------------------------
U46:mux_4to1 PORT MAP(A4=>S15_35,B4=>S15_39,C4=>S15_72,D4=>S15_80,MS4=>SEL_M2,MZ4=>S15_88);
U47:latch2 PORT MAP(L2=>S15_88,LEN2=>X14,W1=>S15_89,W2=>S15_90);

U48:mul PORT MAP(mA=>S15_89,mB=>S15_90,mZ=>S15_91);
U49:latch1 PORT MAP(L1=>S15_91,LEN1=>X15,LZ1=>S15_92);
U50:demux_1to4 PORT MAP(P4=>S15_15,Q4=>S15_18,R4=>S15_57,T4=>S15_63,DS4=>DSEL_M2,DZ4=>S15_92);
```

(i) VHDL register and functional unit port map code snippet

```
-------------------------A2 Block--------------------------------------------------------
U51:mux_8to1 PORT MAP(F0=>S15_34,F1=>S15_42,F2=>S15_46,F3=>S15_71,F4=>S15_75,F5=>S15_79,F6=>S22,F7=>S23,MS8=>SEL_A2,MZ8=>S15_93);

U52:latch2 PORT MAP(L2=>S15_93,LEN2=>X15,W1=>S15_94,W2=>S15_95);

U53:adder PORT MAP(aA=>S15_94,aB=>S15_95,aZ=>S15_96);
U54:latch1 PORT MAP(L1=>S15_96,LEN1=>X16,LZ1=>S15_97);
U55:demux_1to8 PORT MAP(P0=>S15_14,P1=>S15_17,P2=>S15_23,P3=>S15_56,P4=>S15_59,P5=>S15_62,P6=>F1,P7=>F2,DS8=>DSEL_A2,DZ8=>S15_
97);

end structural;
```

(j) VHDL register and functional unit port map code snippet

Figure 6.6 (Continued)

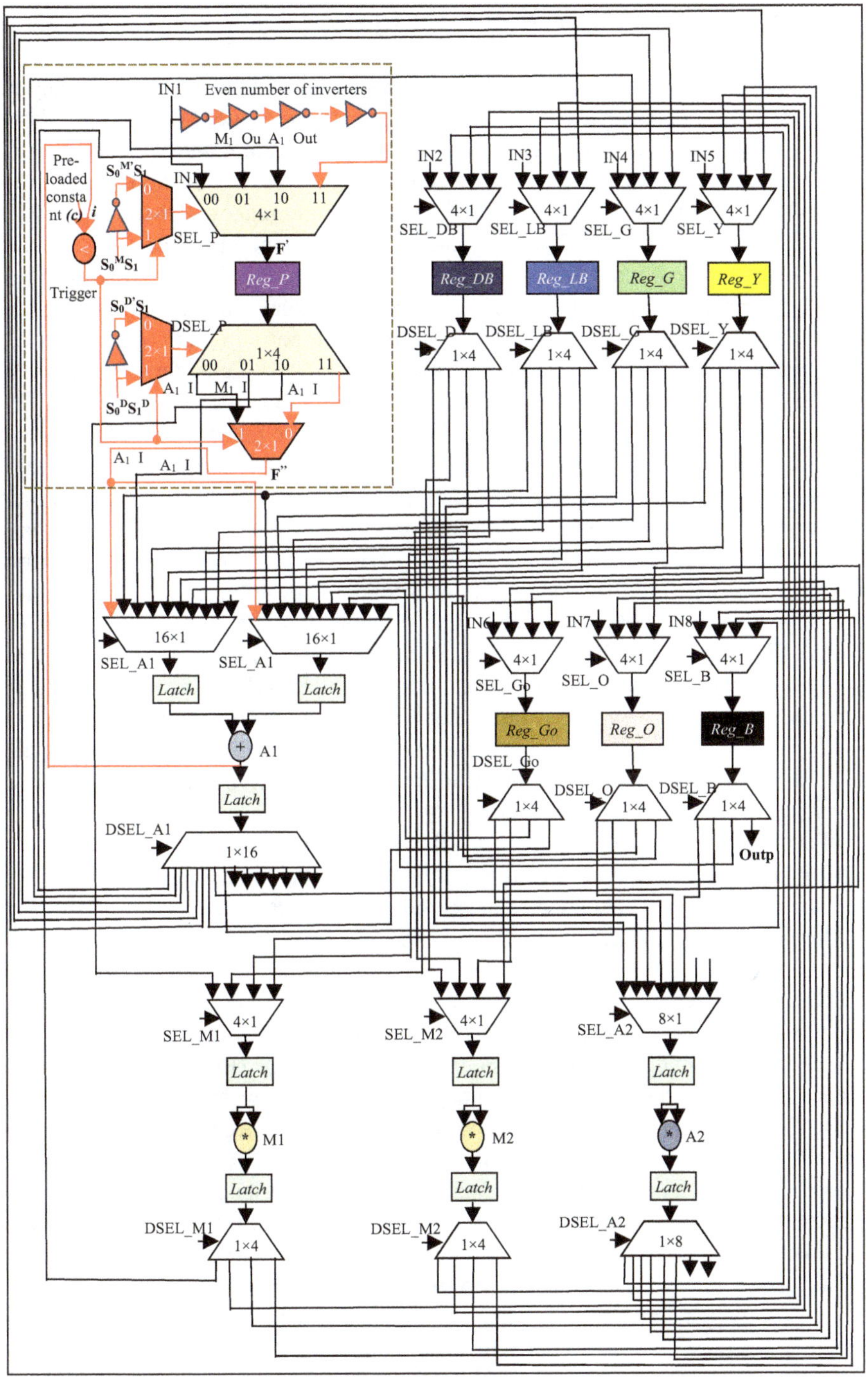

Figure 6.7 Trojan-infected RTL datapath (with PD-HT) corresponding to FIR IP

6.7 Other Trojan detection techniques

Other hardware Trojan detection techniques in the state-of-the-art include Sengupta *et al.* (2017), Abderehman *et al.* (2022), Sengupta *et al.* (2023), Jin and Markis (2008), Ludwig *et al.* (2021), Huang *et al.* (2018), and Rathor and Sengupta (2024), each with different focuses and limitations. Some notable contributions include security-aware scheduling using double module redundancy (DMR) framework (TL-HLS) (Sengupta *et al.*, 2017), C-to-RTL equivalence checking (Abderehman *et al.*, 2022), triple module redundancy (TMR) framework-based Trojan-resilient synthesis methods (Sengupta *et al.*, 2023), path delay fingerprinting (Jin and Markis, 2008), physical inspection and reverse engineering (Ludwig *et al.*, 2021), side-channel analysis-based technique (Huang *et al.*, 2018), and detection based on high-level transformations (HLTs) (Rathor and Sengupta, 2024).

Among these, physical inspection and reverse engineering (Ludwig *et al.*, 2021) are generally impractical in the context of HLS-generated IPs due to their scale and complexity. The RTL derived from HLS is considerably larger and more intricate than the source-level description, making it extremely time-consuming to trace back the design intent. Moreover, many HLS Trojans only activate under rare and carefully designed trigger conditions, meaning the infected IP often behaves normally unless that condition is met. For example, Trojans such as PD-HT, DoS-HT, BE-HT, and DA-HT affect performance, power, or operation/execution without altering the observable data/functionality, while F-HT modifies output data only under specific triggers. These stealthy characteristics make conventional functional validation ineffective. Further, the side-channel-based technique (Huang *et al.*, 2018) is not capable of detection as these Trojans do not leak significant power, timing, or latency information.

Furthermore, frameworks like TL-HLS (Sengupta *et al.*, 2017), which rely on DMR-based scheduling, are effective only for detecting functional Trojans (F-HT) and fail against other Trojan classes. Similarly, C-to-RTL equivalence checking (Abderehman *et al.*, 2022) builds FSMD models from RTL to compare with the input C-code, but it can only capture BE-HT and DA-HT, leaving PD-HT, TB-HT, and DoS-HT undetected. This is because PD-HT, TB-HT, and DoS-HT do not have Trojan-infected C-code and do not exploit the Bambu HLS tool for producing the FSMD RTL module. Therefore, Abderehman *et al.* (2022) cannot apply equivalence checking on these Trojans. Next, path delay fingerprinting (Jin and Markis, 2008), though useful for differentiating infected circuits, becomes infeasible when applied to complex HLS-generated IPs. Similarly, other countermeasures also have limited applicability. HLT-based detection (Rathor and Sengupta, 2024) relies on analyzing operation counts and is only effective against BE-HT, since other Trojan types do not involve inserting redundant operations. Subsequently, the multi-vendor triple modular redundancy strategy in Sengupta *et al.* (2023) provides robustness against functional Trojans but cannot identify other variants.

In short, none of these approaches (Sengupta *et al.*, 2017; Abderehman *et al.*, 2022; Sengupta *et al.*, 2023; Jin and Markis, 2008; Ludwig *et al.*, 2021; Huang

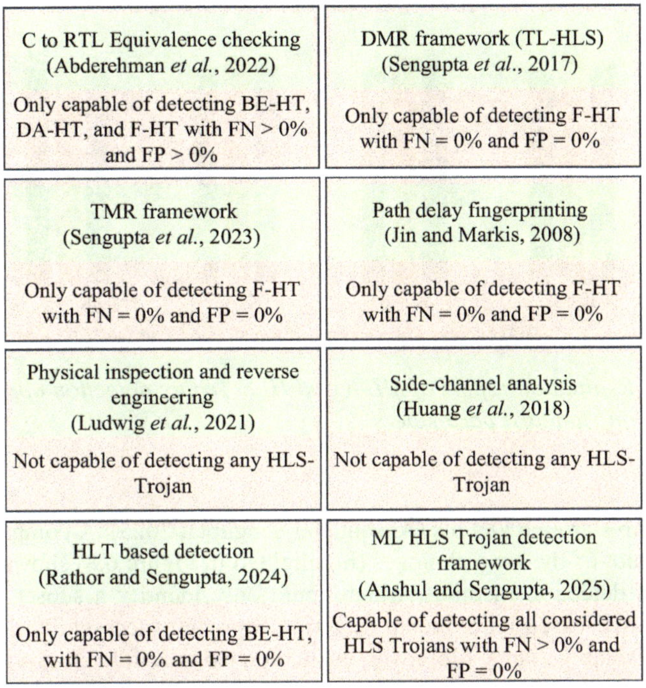

Figure 6.8 Comparison of ML-Trojan detection framework with other state-of-the-art Trojan detection frameworks

et al., 2018; Rathor and Sengupta, 2024) achieves comprehensive coverage of all known HLS Trojan classes. However, the ML-based HLS-Trojan detection methodology (discussed in this chapter) addresses this by employing a custom feature extraction block and decision tree-based machine learning model that achieves complete detection across all Trojan categories (Anshul and Sengupta, 2025). Figure 6.8 highlights a comparative analysis of the ML-Trojan detection framework (Anshul and Sengupta, 2025) with other state-of-the-art Trojan detection frameworks (Sengupta *et al.*, 2017; Abderehman *et al.*, 2022; Sengupta *et al.*, 2023; Jin and Markis, 2008; Ludwig *et al.*, 2021; Huang *et al.*, 2018; Rathor and Sengupta, 2024).

6.8 Analysis of the ML detection technique

This section provides an analytical validation of the discussed ML detection technique. The decision tree classifier (used in the ML detection technique) is implemented using the *scikit-learn* library of the *Python* framework (Anshul and Sengupta, 2025), and the automated feature extraction procedure is implemented in *Python* and executed on the Anaconda platform. The average runtime for the feature extraction process is approximately 19 ms, and the complete Trojan detection

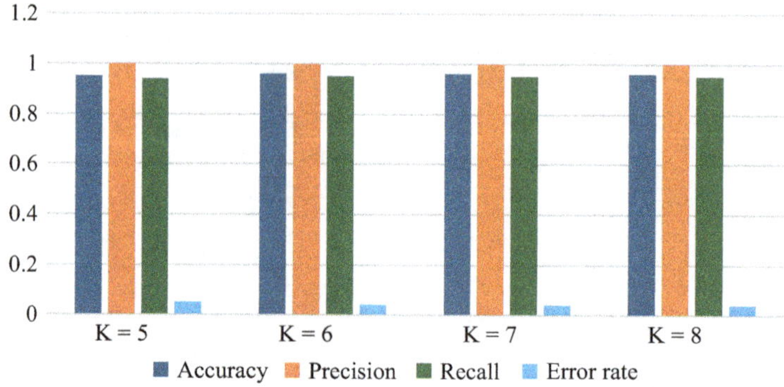

Figure 6.9 Evaluation report of ML-based HLS-Trojan detection approach based on standard parameters

process requires around 789 ms (Anshul and Sengupta, 2025). A comparative study with other state-of-the-art techniques (highlighted in Figure 6.8) shows that, unlike other Trojan detection methods, which could only identify a subset of hardware Trojans, the ML-based HLS-Trojan detection approach successfully detects all six considered Trojan variants. Moreover, evaluation using K-fold cross-validation (for K = 5, 6, 7, 8) demonstrates an average accuracy in the range of 95–96%, with precision values of 1.0 and recall values around 0.94–0.95, indicating balanced and reliable classification performance (with a high detection rate). Figure 6.9 illustrates the evaluation report of the ML-based HLS-Trojan detection approach based on standard parameters for different K-values. The error rate remained low (4–5%), validating the effectiveness of the decision tree-based ML detection methodology.

6.9 Conclusion

An HLS-Trojan detection framework has been discussed in this chapter that employs a machine learning classifier to distinguish between hardware IP RTL designs that are compromised by Trojans and those that are clean. A custom feature extraction block has also been discussed, which extracts structural and behavioral attributes from the given input RTL design, which are then provided as inputs to the ML classification model for training and testing purposes. The analytical validation demonstrated that the discussed ML-based HLS-Trojan detection methodology achieves a high detection rate for all considered Trojans.

6.10 Questions and exercises

1. Explain machine learning and its applications in brief.
2. List down the types of HLS Trojans in the state-of-the-art.

3. Why machine learning is important in the context of HLS-Trojan detection.
4. Write a short note on the threat model discussed in this chapter in the context of the hardware Trojan attack.
5. Briefly explain the impact of the different HLS Trojans discussed in this chapter.
6. Which machine learning classifier has been used to perform Trojan detection, and why?
7. Discuss the custom feature extraction block discussed in this chapter and its importance.
8. Draw the complete flow diagram of the ML-based HLS-Trojan detection methodology and explain each step in detail.
9. Explain the advantages of the discussed ML-based HLS-Trojan detection approach over other detection approaches in the state-of-the-art.
10. Implement the feature extraction module (in your preferred language) and compare the feature extraction time with the one discussed.

References

T. Hastie, R. Tibshirani, and J. Friedman, (2009), *The Elements of Statistical Learning: Data Mining, Inference, and Prediction*, 2nd ed. New York, NY, USA: Springer, 2009.

C. Bishop, (2006), *Pattern Recognition and Machine Learning*. Springer, 2006.

R. S. Sutton, and A. G. Barto, (2018), *Reinforcement Learning: An Introduction*, 2nd ed. MIT Press, 2018.

M. Rostami, F. Koushanfar, and R. Karri, (2014), "A primer on hardware security: models, methods, and metrics," *Proceedings of the IEEE*, vol. 102, no. 8, pp. 1283–1295, 2014.

A. Anshul, and A. Sengupta, (2025), "HLS Trojan detection using machine learning technique," *IEEE Embedded Systems Letters*, 2025, doi:10.1109/LES.2025.3614187.

Y. Jin, and Y. Makris, (2008), "Hardware Trojan detection using path delay fingerprint," *Proceedings of the IEEE International Workshop on Hardware-Oriented Security and Trust (HOST)*, 2008, pp. 51–57.

H. Pearce, R. Karri, and B. Tan, (2023), "High-level approaches to hardware security: a tutorial," *ACM Transactions on Embedded Computing Systems*, vol. 22, no. 3, article 45, 2023.

A. Sengupta, A. Anshul, V. Chourasia, and N. Kumar, (2024), "M-HLS: malevolent high-level synthesis for watermarked hardware IPs," *IEEE Embedded Systems Letters*, vol. 16, no. 4, pp. 497–500, 2024.

C. Pilato, K. Basu, F. Regazzoni, and R. Karri, (2019), "Black-Hat high-level synthesis: myth or reality?," *IEEE Transactions on Very Large Scale Integration (VLSI) Systems*, vol. 27, no. 4, pp. 913–926, 2019.

A. Sengupta, S. Bhadauria, and S. P. Mohanty, (2017), "TL-HLS: methodology for low cost hardware Trojan security aware scheduling with optimal loop unrolling factor during high level synthesis," *IEEE Transactions on Computer-Aided Design of Integrated Circuits and Systems*, vol. 36, no. 4, pp. 655–668, 2017.

A. Sengupta, and N. Bhui, (2025), "Time-bomb HLS Trojan for performance degradation payload," *IEEE Design & Test*, vol. 42, no. 5, pp. 83–92, 2025.

A. Sengupta, A. Anshul, V. Chourasia, and N. Bhui, (2025), "Security vulnerability (Backdoor Trojan) during machine learning accelerator design phases," *IT Professional*, vol. 27, no. 1, pp. 65–72, 2025.

A. Anshul, and A. Sengupta, (2023a), "A survey of high level synthesis based hardware security approaches for reusable IP cores," *IEEE Circuits and Systems Magazine*, vol. 23, no. 4, pp. 44–62, Fourthquarter 2023.

M. Rathor, A. Anshul, and A. Sengupta, (2014), "Securing reusable IP cores using voice biometric based watermark," *IEEE Transactions on Dependable and Secure Computing*, vol. 21, no. 4, pp. 2735–2749, 2024.

A. Anshul, and A. Sengupta, (2023b), Exploration of optimal crypto-chain signature embedded secure JPEG-CODEC hardware IP during high level synthesis, *Microprocessors and Microsystems*, Vol 102, pp. 104916, 2023.

GitHub Repository, (2025), Available [Online]: https://github.com/ryderaadi/HLS-Trojan-Database–HLS-TD.

M. Abderehman, R. Gupta, R. R. Theegala, and C. Karfa, (2022), "BLAST: belling the Black-Hat high-level synthesis tool," *IEEE Transactions on Computer-Aided Design of Integrated Circuits and Systems*, vol. 41, no. 11, pp. 3661–3672, 2022.

A. Sengupta, A. Anshul, and R. Chaurasia, (2023), Exploration of optimal functional Trojan-resistant hardware intellectual property (IP) core designs during high level synthesis, *Elsevier Microprocessors and Microsystems*, vol. 103, 2023, pp. 104973, 2023.

M. Ludwig, A. -C. Bette, and B. Lippmann, (2021), "ViTaL: verifying Trojan-free physical layouts through hardware reverse engineering," *IEEE Physical Assurance and Inspection of Electronics*, USA, 2021, pp. 1–8.

Y. Huang, S. Bhunia, and P. Mishra, (2018), "Scalable test generation for Trojan detection using side channel analysis," *IEEE Transactions on Information Forensics and Security*, vol. 13, no. 11, pp. 2746–2760, 2018.

M. Rathor, and A. Sengupta, (2024), "Revisiting Black-Hat HLS: a lightweight countermeasure to HLS-aided Trojan attack," *IEEE Embedded Systems Letters*, Volume: 16, Issue: 2, 2024, pp. 170–173.

Appendix

Python code of feature extraction:

```
import re
import pandas as pd
start = time_1.time()
with open ("Datapath VHDL_with portmap.txt", 'r') as file:
```

```
    content = file.readlines()
# def entity_Presence(lst):
#     look = False
#     unwanted = ['‚','‚';','(', ')', ':']
#     for line in content:
#         if 'entity' in line:
#             look=True
#         if look:

r"mux(?:[1-9]|1[0-9]|20)_(?:2|4|8|16|32)_to_(?:1|2|4|8|16)"
def muxCount(type):
    muxCnt = 0
    pattern = rf"component mux(?:[0-9]|1[0-9]|20)?_{type}_to_1"
    print(pattern)
    for line in content:
        if re.search(pattern, line):
        print(line)
        muxCnt+=1

    return muxCnt

def deMuxCount(type):
    deMuxCnt = 0
    pattern = rf"component ([d|D]emux)(?:[0-9]|1[0-9]|20)?_{type}_to_1"
    for line in content:
        if re.search(pattern, line):
            print(line)
            deMuxCnt+=1

    return deMuxCnt

def notGateCount():
    pattern = r"component ([N|n]ot)(?:[1-9]|1[0-9]|2[0-5])\b"
    checkNot = False
    compDefnList = {}
    itr = 0
    container = {'input':[], 'output':[]}
    defn = ""
    for line in content:
        if re.search(pattern, line) and not checkNot:
            checkNot=True
            res = line.split(" ")
            defn += res[1]
        if checkNot and re.search(r'end component', line):
            itr = 0
            compDefnList[defn]=container
            container={'input':[], 'output':[]}
```

```
            defn = ""
            checkNot=False
        if checkNot:
            res = re.split(r'[;:,]', line)
            for i in range(res):
                if res[i]=='in':
                container['input'].append(itr)
                itr+=1
        if res[i]=='out':
                container['output'].append(itr)
                itr+=1
        notIp,notOp = [],[]
        for key,value in compDefnList:
            notIp.extend(i for i in value['input'])
            notOp.extend(i for i in value['output'])
        notChain = False
        for i in notIp:
            if i in notOp:
                notChain=True
        print('not chain present' if notChain else ")
        return len(compDefnList),notChain

    def counterCount():
        counterCnt = 0
        pattern = r"component counter(?:[1-9]|1[0-9]|2[0-5])\b"
        for line in content:
            if re.search(pattern, line):
            counterCnt+=1
        return counterCnt

    def comparaterCount():
        comperatorCnt = 0
        pattern = r"component comparator(?:[1-9]|1[0-9]|2[0-5])\b"
        for line in content:
            if re.search(pattern, line):
            comparaterCount+=1
        if comperatorCnt>=1:
            print("check for extra register ")
    else:
        print("dont check for extra register ")
    return comperatorCnt, True if comperatorCnt else False

    def triStateBuffer():
        TSBCnt = 0
        pattern = r"component tri-state buffer(?:[1-9]|1[0-9]|2[0-5])\b"
        for line in content:
```

```
    if re.search(pattern, line):
        TSBCnt+=1
    return TSBCnt

# compCount, presence = comparaterCount()
def extraRegister(presence):
    # presence=True
    regOutput = []
    regDefnList = {}
    pattern = r"component comparator(?:[1-9]|1[0-9]|2[0-5])\b"
    regPattern = r"component register[A-Z]{1,2}"
    checkComp, checkReg = False, False
    extraReg = 0
    compDefnList = {}
    itr = 0
    container = {'input':[], 'output':[]}
    defn = ""
    check=False
    for line in content:
        if re.search(pattern, line) and not checkComp:
            checkComp=True
            res = line.split(" ")
            defn += res[1]
        if checkComp and re.search(r'end component', line):
            itr = 0
            compDefnList[defn]=container
            container={'input':[], 'output':[]}
            defn = ""
            checkComp=False
        if checkComp:
            res = re.split(r'[;:,]', line)
            for i in range(res):
                if res[i]=='in':
                    container['input'].append(itr)
                    itr+=1
                if res[i]=='out':
                    container['output'].append(itr)
                    itr+=1
        if re.search(regPattern, line) and not checkReg:
            checkReg=True
            res=line.split()
            defn+=res[1]
        if checkReg and re.search(r'end component', line):
            itr = 0
```

```
            regDefnList[defn]=container
            container={'input':[], 'output':[]}
            defn = ""
            checkReg=False
        if checkReg:
            res = re.split(r'[;:, ]', line)
            if " in res:
                res.remove(")
            for i in range(len(res)):
                if res[i]=='in':
                    if 'input' not in container:
                        container['input']=[]
                    container['input'].append(itr)
                    itr+=1
                if res[i]=='out':
                    container['output'].append(itr)
                    itr+=1

    if presence:
        for line in content:
            if 'begin' in line:
                check=True
    if check and re.search(r"register[A-Z]{1,2}", line):
        print(line)
        res = re.split(r'[: ]', line)
        if " in res:
            res.remove(")
        defn=res[1]
        match = re.search(r"\((.*?)\)", line)
        inside_brackets = match.group(1)
        elements = [elem.strip() for elem in inside_brackets.split(",")]
regOutput.extend([elements[i] for i in regDefnList[defn]['output']])
    print(regOutput)
    check = False
    for line in content:
        if 'begin' in line:
            check=True
        if check and re.search(r'comparator(?:[1-9]|1[0-9]|2[0-5])\b', line):
            res.split = re.split(r'[: ]',line)
            if " in res:
                res.remove(")
            defn=res[1]
            match = re.search(r"\((.*?)\)", line)
            inside_brackets = match.group(1)
            elements = [elem.strip() for elem in inside_brackets.split(",")]
```

```
        print(compDefnList[defn]['input'])
        for i in compDefnList[defn]['input']:
            if elements[i] in regOutput:
            extraReg+=1
    if extraReg:
        print('extra register present!')
    return extraReg, regOutput, len(regDefnList)
return 0,regOutput, len(regDefnList)

def muxToMux():
    pattern = r"component (?<!de)mux(?:[0-9]|1[0-9]|20)?_(?:2|4|8|16|32)_to_(?:1)"
    checkMux = False
    regDefnList={}
    container={'output':[],'input':[]}
    defn=""
    itr=0
    for line in content:
        if re.search(pattern, line) and not checkMux:
            checkMux=True
            res = line.split(" ")
            defn += res[1]
        if checkMux and re.search(r'end component', line):
            # print(line)
            itr = 0
            regDefnList[defn]=container
            container={'output':[],'input':[]}
            defn = ""
            checkMux=False
        if checkMux:
            res = re.split(r'[;:, ]', line)
            res = [item for item in res if item]
            # print(res)
            for i in range(len(res)):
                if res[i]=='in':
                    container['input'].append(itr)
                    itr+=1
                if res[i]=='out':
                    container['output'].append(itr)
                    itr+=1
            # print(regDefnList)
pattern = r"(?<![D|d]e)mux(?:[0-9]|1[0-9]|20)?_(?:2|4|8|16|32)_to_(?:1|2|4|8|16)"
    muxInput = []
    defn=""
    check=False
    for line in content:
```

```
if 'begin' in line:
   check=True
if check and re.search(pattern, line):
   print(line)
   res = re.split(r'[: ]', line)
   res = [item for item in res if item]
   # print(res)
   defn=res[1]
   match = re.search(r"\((.*?)\)", line)
   if match:
      inside_brackets = match.group(1)
      elements = [elem.strip() for elem in inside_brackets.split(",")]
      # print(defn)
      for i in regDefnList[defn]['input']:
         muxInput.append(elements[i])
# print(muxInput)
mTmCnt = 0
defn=""
for line in content:
   if 'begin' in line:
      check=True
   if check and re.search(pattern, line):
      # print(line)
      res = re.split(r'[: ]', line)
      res = [item for item in res if item]
      defn=res[1]
      match = re.search(r"\((.*?)\)", line)
      if match:
         inside_brackets = match.group(1)
         elements = [elem.strip() for elem in inside_brackets.split(",")]
         for i in regDefnList[defn]['output']:
            # print(elements[i],sep=" ")
            if elements[i] in muxInput:
               mTmCnt+=1
if mTmCnt>0:
   print("mux to mux entry present!")
return mTmCnt, muxInput
def functionalUnit():
   check = False
      pattern = r"component ([A|a]dder|[M|m]ultiplier)(?:[0-9]|1[0-9]|20)?"
      patt = r"\b(\w+)\b(?=\s*:\s*(in|out))"
      present = []
      for line in content:
         if "component control_unit" in line and not check:
            check=True
```

```
            elif check and re.search(r"end component", line):
                check=False
            elif check:
                # print(line)
                res = re.split(r'[;:, ]', line)
        res = [item for item in res if item]
        print(res)
        present.extend([i for i in res])
    unwanted_strings = {"\n", "in", "out", "std_logic", "INTEGER", "port(", ")", ""}

    # Filter the list to keep only valid signal names
    present = [
        item for item in present
        if item not in unwanted_strings and not item.startswith("–") and not item.
isspace()
    ]
    check=False
    trojan = False
    for line in content:
        if re.search(pattern, line) and not check:
            print(line)
            check=True
        if check and re.search("end component", line):
            check=False
        if check:
            res = re.split(r'[;:, ]', line)
            res = [item for item in res if item]
            print(res)
            for i in range(len(res)):
                if res[i]=='in' and 'enable' in res[i-1]:
                print(res[i-1], res[i])
                if res[i-1] not in present:
                    trojan=True

    print('Trojan Present' if trojan else None)
    return trojan

  def regToMux(regOp, muxIp):
    extraReg = 0
    for i in regOp:
        if i in muxIp:
            extraReg+=1
    #print("extra Register Present")
    return extraReg
    def latch():
        latchCnt = 0
```

```
        pattern = r"component ([D|T]_[l|L]atch)(?:[1-9]|1[0-9]|2[0-5])"
        for line in content:
            if re.search(pattern, line):
                print(line)
                latchCnt+=1
        return latchCnt

def FU(s):
    Cnt = 0
    pattern = rf"component ([{s[0]}|{s[0].upper()}]{s[1:]})(?:[0-9]|1[0-9]|20)?"
    print(pattern)
    for line in content:
        if re.search(pattern, line):
        Cnt+=1
    return Cnt
if __name__ == "__main__":
    featureSet = {'features':[],'present':[],'count':[]}
    i,itr = 2,1
    while i<=32:
        mxCnt = muxCount(i)
        featureSet['features'].append(f'mux{i}_to_1')
        featureSet['present'].append('Yes' if mxCnt>0 else 'No')
        featureSet['count'].append(mxCnt)
        itr+=1
        i = pow(2,itr)

    i,itr=2,1
    while i<=32:
        mxCnt = deMuxCount(i)
        featureSet['features'].append(f'demux{i}_to_1')
        featureSet['present'].append('Yes' if mxCnt>0 else 'No')
        featureSet['count'].append(mxCnt)
        itr+=1
        i = pow(2,itr)
    notCnt, notChain = notGateCount()
    featureSet['features'].append('not gate')
    featureSet['present'].append('Yes' if notCnt>0 else 'No')
    featureSet['count'].append(notCnt)
    counterCnt = counterCount()
    featureSet['features'].append('counter')
    featureSet['present'].append('Yes' if counterCnt>0 else 'No')
    featureSet['count'].append(counterCnt)
    TSBCnt = triStateBuffer()
    featureSet['features'].append('Tri State Buffer')
    featureSet['present'].append('Yes' if TSBCnt>0 else 'No')
    featureSet['count'].append(TSBCnt)
```

```
compCount, presence = comparaterCount()
featureSet['features'].append('Comparator')
featureSet['present'].append('Yes' if compCount>0 else 'No')
featureSet['count'].append(compCount)
extra, regOp, regCnt = extraRegister(presence)
featureSet['features'].append('registers')
featureSet['present'].append('Yes' if regCnt>0 else 'No')
featureSet['count'].append(regCnt)

mtmPresent, muxIp = muxToMux()
extraReg = regToMux(regOp, muxIp)

latchCnt = latch()
featureSet['features'].append('latch')
featureSet['present'].append('Yes' if latchCnt>0 else 'No')
featureSet['count'].append(latchCnt)

fnUnit = functionalUnit()

Adder = FU('adder')
featureSet['features'].append('Adder')
featureSet['present'].append('Yes' if Adder>0 else 'No')
featureSet['count'].append(Adder)
Subtractor = FU('subtractor')
featureSet['features'].append('Subtractor')
featureSet['present'].append('Yes' if Subtractor>0 else 'No')
featureSet['count'].append(Subtractor)
multiplier = FU('multiplier')
featureSet['features'].append('Multiplier')
featureSet['present'].append('Yes' if multiplier>0 else 'No')
featureSet['count'].append(multiplier)
df = pd.DataFrame(featureSet)
print(df)
```

Chapter 7

HLS Trojan detection using Dual Modular Redundancy (DMR) techniques

Anirban Sengupta[1]

The chapter provides the readers with an introduction to the increasing reliance on micro-third-party intellectual properties (IPs) used in IP design/system-on-chip (SoC) design, which has raised concerns about their security and integrity. Malicious logic, commonly known as hardware Trojans, can be covertly inserted into these third-party components used in the high-level synthesis (HLS) tool library. These HLS Trojans are called functional Trojans, as they affect the functional computational output of IP design. This chapter focuses on the detection of such Trojans during the high-level synthesis phase of digital IP design, where third-party modules are integrated. Implementing detection mechanisms for such threats requires additional hardware, which could lead to violations of area and timing constraints. The challenge lies in balancing Trojan detection with design constraints, as incorporating extra logic for detection may impact area overhead. This chapter explores the trade-offs involved in ensuring Trojan-secure IP design while integrating the design space exploration framework during HLS.

The rest of the chapter is structured as follows: Section 7.1 provides the introduction of the chapter; Section 7.2 outlines salient features of chapter; Section 7.3 presents brief discussion on the threat model; Section 7.4 defines HLS functional Trojans; Section 7.5 presents the discussion of different Trojan detection techniques; Section 7.6 highlights the need for security-centric HLS; Section 7.7 discusses the details of the Trojan security-aware HLS process; Section 7.8 describes the PSO-driven exploration of the Trojan-secured datapath; Section 7.9 provides details of designing an optimal Trojan-secured design schedule using design space exploration (DSE); Section 7.10 discusses the terminating criteria of the process; Section 7.11 presents an analysis of case studies; and Section 7.12 concludes the chapter by summarizing its significance and implications.

[1]Department of Computer Science and Engineering, Indian Institute of Technology Indore, India

7.1 Introduction

In the context of modern integrated circuit (IC) design, hardware Trojan (HT) components refer to malicious logic or hardware elements inserted through backdoor entry points. The HLS-based HT are deliberately introduced into a system by a rogue actor with the intent to cause malfunction, misbehavior, or to extract sensitive information from the IC. Similarly, these Trojans can manifest in various forms of threats, from subtle bugs in the design to more aggressive, hard-to-detect alterations of the hardware that can lead to catastrophic system failures or data breaches. Such threats are particularly alarming in the realm of System-on-Chip (SoC) design, where millions of transistors and complex interactions between analog and digital components are packed into a single chip. With the increasing globalization of SoC design and manufacturing, there are growing concerns over the security and trustworthiness of the embedded third-party intellectual property (3PIP) (Sengupta *et al.*, 2017; Zhang and Tehranipoor, 2011; Banga and Hsiao, 2009; Karri *et al.*, 2010; Bhunia *et al.*, 2013). Third-party IP refers to components, modules, or libraries that are purchased from external (different) IP vendors for integration into the design of the SoC. While outsourcing various components can offer significant advantages in terms of cost, time, and access to specialized technologies, it also opens the door to potential security vulnerabilities.

A typical SoC designing and manufacturing cycle involves multiple stages and often a combination of internal and external contributors, as shown in Figure 7.1. The

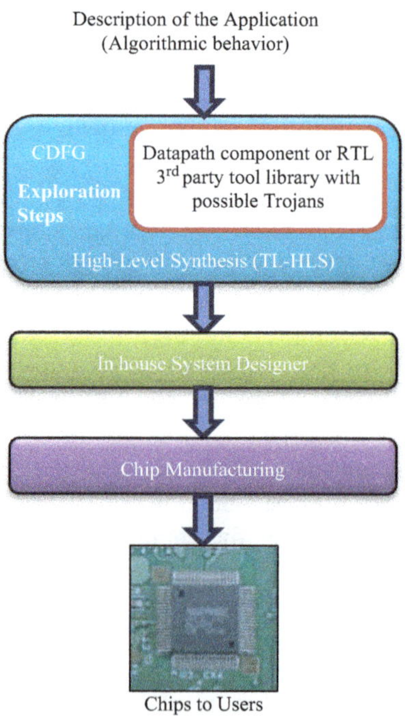

Figure 7.1 Potential hardware Trojan in the HLS tool library during chip development

designing and manufacturing process of a digital integrated circuit begins with a primary input that is fundamentally a detailed and precise description of the application (algorithm) to be implemented using HLS. This description typically takes the form of high-level design specifications, which may include behavioral models, control data flow graph (CDFG), and C/C++/hardware description language (HDL) code. These descriptions encapsulate the intended functionality, timing requirements, data flow, and control logic of the digital system to be realized. The final output of this extensive and complex process is the actual silicon user chip, a physical piece of hardware that is the output. These chips, as part of end products, are delivered to users, whether individuals, corporations, or embedded within larger electronic systems such as smartphones, computers, automotive controls, or industrial machinery.

As shown in Figure 7.1, one of the critical stages is the inclusion of third-party datapath components, typically represented as register transfer level (RTL) libraries in the HLS tools. These library components are part of the system's architecture, controlling the flow of data through the datapath. However, these third-party libraries may be considered untrustworthy, as their integrity cannot be guaranteed. In contrast, the rest of the SoC design process carried out by trusted in-house designers is generally assumed to be secure. This trust model becomes particularly problematic when these third-party components are not rigorously vetted for security vulnerabilities or potential malicious logic. A hardware HLS Trojan could be covertly embedded into the HLS design process using third-party IP in the tool library. Unless proper security mechanisms are in place, it could remain hidden and active during the HLS design process and could end up being part of the final chip.

Moreover, SoCs are often designed to combine both analog and digital components in order to optimize the trade-off between performance and cost (Sengupta *et al.,* 2017; Mohanty, 2015). These mixed-signal designs allow for more efficient systems by incorporating specialized functions like analog-to-digital conversion, radio frequency processing, and power management alongside the digital logic that governs computation and control tasks. While this integration provides significant benefits, it also introduces additional complexity and increases the attack surface for potential security breaches.

The primary focus of this chapter, however, is on the digital components of SoCs, particularly when these components undergo design exploration at the architecture level. At this stage, the architecture and overall functionality of the system are defined, and decisions are made regarding the structure of the datapath, control logic, and interfaces between different system blocks. While the focus may primarily be on performance optimization, power efficiency, and cost, security considerations, especially the risk of hardware Trojans, must also be incorporated at this crucial phase to mitigate potential threats in the final product.

In the following sections, we will explore how hardware HLS Trojans are introduced at different stages of the SoC design process, their potential impact on system performance and security, and strategies for mitigating these risks through improved design practices, verification techniques, and trusted third-party IP sourcing. In the modern landscape of IC and SoC design, the use of 3PIP has become increasingly common. These third-party modules or components, which are typically part of a tool's module library in the HLS process, provide substantial benefits by reducing development time and costs. However, this practice introduces significant security risks, primarily due to the potential for malicious alterations, such as hardware Trojans, which could be inserted into the design by an adversary.

A hardware Trojan is a malicious modification to an integrated circuit, often introduced during the design phase, with the intent to compromise the functionality, security, or performance of the final product. These Trojans can be inserted covertly, often in a manner that is difficult to detect, and they can have devastating effects once the hardware is deployed in the field. Given that third-party IP modules may be imported from external vendors, their security cannot always be guaranteed. Consequently, a critical challenge in SoC design is ensuring that any potential Trojan logic hidden within a third-party module is detected early, particularly during the HLS phase of design (Sengupta *et al.*, 2017; Rajendran *et al.*, 2013).

HLS is a process in which a higher-level description of the system, usually in the form of a high-level programming language or algorithm, is converted into a lower-level hardware description, typically in RTL code. It is during this phase that the various components of the design, including any third-party modules, are synthesized into a hardware blueprint. However, this is also a phase where vulnerabilities can be introduced, particularly if the modules imported from third-party vendors, as library components of the HLS tool, have been compromised as discussed above.

An HLS tool provides an RTL library consisting of pre-designed modules or micro third-party IPs, some of which may have been sourced from third-party vendors. While these micro-IPs are often trusted by the tool user, they are, in fact, untrustworthy by default. The third-party vendor introduces a security risk in the form of backdoor Trojan logic in such micro-IPs. The potential for such insertion is alarming because even subtle, hidden changes to the HLS-generated RTL code could lead to catastrophic failures (Sengupta *et al.*, 2025).

As highlighted in state-of-the-art (Sengupta *et al.*, 2017; Rajendran *et al.*, 2013), this risk is particularly concerning because during the HLS phase, the imported third-party modules may contain hardware Trojan logic, which can remain undetected unless specific detection strategies are employed. These HLS hardware Trojans might manipulate the data path and induce functional error. The RTL file, which is formed using these modules, might contain such Trojan logic. Therefore, unless proactive detection mechanisms are put in place, these malicious alterations could escape notice during the HLS design process (Sengupta *et al.*, 2025, 2024).

Given the significant risk of hardware Trojan insertion in the HLS tool library, the design process during HLS must include a detection strategy to identify potential threats in third-party modules (micro-IP). The detection method must aim to ensure that any malicious modification introduced into the HLS tool library can be detected and addressed. However, the detection of hardware Trojans during the HLS IP design process presents a significant challenge for several reasons. One of the primary concerns is the additional hardware that may be required for detection. Implementing a Trojan detection mechanism often necessitates the inclusion of extra logic in the design, which, when deployed, could exceed the area constraints specified by the user. This means that the design might no longer fit within the allocated hardware area, leading to increased cost.

To address these challenges, it becomes essential to consider the effects of extra delay/latency and hardware cost during design space exploration (DSE) in HLS. DSE refers to the process of exploring various design alternatives to optimize key performance metrics, such as area, power consumption, timing, and cost. When incorporating hardware Trojan detection logic, the HLS-based IP designer must evaluate how the extra hardware/resource overhead and delay/latency impact the overall design's feasibility within the given constraints. This process of exploration involves multi-objective optimization, where designers must balance security requirements, ensuring that hardware Trojans are detected and mitigated, with the traditional design goals of performance, area, and power. In particular, designers need to assess whether the hardware Trojan detection mechanisms can be implemented without violating the designer-defined area and delay constraints. This may require iterative refinement of the IP design, adjusting both the Trojan detection strategy and the core architecture of the IP design to maintain a balance between security and design cost (Sengupta *et al.*, 2017).

The design and synthesis processes described earlier primarily apply to data flow graphs (DFGs) within HLS. DFGs represent the flow of data through a system without explicit control flow constructs, making scheduling and resource allocation relatively straightforward. However, when considering loop-based CDFGs, which introduce iterative control structures such as loops, the design complexity increases significantly. This is due to the presence of an additional crucial parameter, the loop unrolling factor. Loop unrolling is a well-known optimization technique in digital circuit design and HLS, where iterations of a loop are replicated multiple times to increase parallelism, thereby potentially improving performance by reducing execution time. However, this improvement comes at the expense of increased hardware area, as unrolling expands the number of operations that must be executed in parallel. Hence, the loop unrolling factor has a profound impact on the area-delay trade-off of a design; it directly influences how large the circuit will be and how quickly it can process data.

Incorporating hardware Trojan security awareness into the scheduling of CDFGs presents a unique challenge. Simply generating a low-cost Trojan-secure schedule for CDFGs is not sufficient. Unlike traditional DFG scheduling, which focuses on minimizing hardware cost and meeting latency/delay constraints, CDFG scheduling must also take into account how loop unrolling affects the overall area and delay of the design. Neglecting this factor could result in schedules that are either too large or too slow to meet designer constraints, even if they offer Trojan security.

Therefore, an effective design strategy for Trojan-aware scheduling in loop-based control data flow graphs must simultaneously consider:

- The Trojan security requirements, often implemented via redundancy, detect malicious modification but may introduce overhead.
- The optimal loop unrolling factor, which impacts both the resource utilization and timing performance of the design.

To address this complex interplay, this chapter discusses a comprehensive methodology that performs simultaneous design space exploration (DSE) of:

1. A low-cost Trojan security-aware scheduling technique based on Dual Modular Redundancy (DMR), and
2. The optimal loop unrolling factor that satisfies the given area-delay constraints specified by the designer.

DMR is a widely used technique to detect and mitigate hardware Trojan attacks by duplicating operations and comparing their outputs for inconsistencies. While effective in enhancing security and reliability, DMR naturally increases resource usage because it requires doubling the functional units and additional logic to manage and compare outputs. However, the state-of-the-art carefully designs the Trojan-secured DMR schedule to avoid unnecessary overhead by operating within resource constraints derived through an iterative optimization process. This ensures that the design maintains a low-cost profile while providing robust Trojan security (Sengupta *et al.*, 2017).

A key contribution in this state-of-the-art is the use of Particle Swarm Optimization (PSO), which is a bio-inspired, population-based optimization technique to iteratively search the design space. PSO intelligently navigates trade-offs between resource usage and performance, refining both the DMR scheduling and the loop unrolling factor simultaneously to find an optimal solution.

This process results in a Trojan-secure schedule that:

- Minimizes hardware resource duplication,
- Controls the overhead introduced by steering logic and internal buffers necessary for managing redundant operations,
- Meets or exceeds the user-specified constraints on area and delay,
- And maintains functional correctness.

A notable aspect of the state-of-the-art is the careful attention paid to reducing the additional hardware overheads introduced by Trojan security measures. Beyond simply duplicating operations, DMR necessitates steering logic control circuitry to route data correctly between modules and internal buffering to synchronize duplicated operations. These elements can contribute significantly to area and timing penalties if not managed properly. The design technique optimizes these overheads by integrating their costs into the PSO-based exploration, ensuring that the resulting design is efficient not only in terms of core computational resources but also in supporting hardware required for redundancy and security (Mishra and Sengupta, 2014).

7.2 Salient features of the chapter

This chapter presents the state-of-the-art on designing trusted hardware IP, exploiting HLS-based low-cost Trojan detection using DMR scheduling, vendor allocation, and area-delay optimization. Detailed explanation of the points is outlined as follows:

- This chapter explains how to explore a low-cost Trojan-secured schedule (capable of HLS functional Trojan detection) while optimizing both DMR scheduling and the optimal loop unrolling factor during HLS. It explains how to meet designer-defined area-delay constraints while ensuring a Trojan-secure design process. Unlike the Triple Modular Redundancy (TMR) technique (Cui *et al.*, 2014; McIntyre *et al.*, 2009), which focuses on recovery techniques, this chapter exclusively addresses low-cost HLS Trojan detection, rather than system recovery after Trojan activation. Additionally, the focus is on low-cost, Trojan-secure DMR schedules and not on designing Triple Modular Redundancy (TMR) logic.

- This chapter also explains a distinct IP vendor allocation exploration procedure that leverages a PSO-driven DSE process for HLS Trojan detection. The PSO-based encoding identifies the most efficient vendor assignments to achieve Trojan detection, thereby optimizing the secure hardware IP designs.
- A unique encoding scheme is discussed, which represents candidate schedule resources for task scheduling aimed at enhancing security, a candidate loop unrolling factor to improve performance, and candidate vendor allocation information that assigns resources to IP vendors to achieve Trojan detection. This encoding facilitates the simultaneous consideration of all key factors influencing the security of IP designs during the HLS tool library.
- This chapter further explores the area-delay trade-off optimization using PSO during the Trojan-secure IP design process based on DMR logic. DMR inherently requires duplication of CDFG, resulting in additional steering logic and internal buffering compared to non-redundant designs. This chapter demonstrates integration of low-cost Trojan security with DMR scheduling and PSO-DSE, thereby causing minimal resource overhead. This is in contrast to (Rajendran *et al.*, 2013), which often results in higher resource consumption. This chapter describes how to adhere to resource constraints specified through the PSO process, effectively minimizing unnecessary overhead, while still ensuring robust Trojan detection.
- Additionally, this chapter discusses a delay estimation model tailored for a Trojan-secured DMR-CDFG. This model provides accurate timing and delay estimates, which are critical for understanding the performance implications of incorporating a Trojan detection mechanism within the DMR design. By enabling precise delay predictions, the model supports designers in making informed trade-offs between security and performance during the high-level synthesis process.

7.3 Discussion on the threat model

This chapter focuses on a Trojan covertly inserted in the micro-3PIP (as part of the HLS tool library) that can alter computational output value (functional change) without causing other payloads. This chapter does not discuss Trojans designed to steal information or data. The insertion of such Trojans typically occurs when a malicious module or micro-IP is secretly inserted into the library of an HLS tool, either by a rogue individual in the third-party IP design house. Such Trojans are only activated under rare event conditions, therefore it becomes challenging for the validation team to detect their presence.

This chapter assumes that both the foundry and the in-house system designer are trustworthy within the IC design flow. In contrast, the third-party IP vendor, specifically the HLS tool developer, is regarded as untrustworthy. This means that the potential for manipulation of the IP originates solely from the third-party source and the third-party HLS tool. As the entire set of IPs is designed using a third-party HLS tool, the IPs may inadvertently contain malicious Trojan logic. Figures 7.2–7.4 show compromised (Trojan-infected) micro 3PIPs as modules/components in the

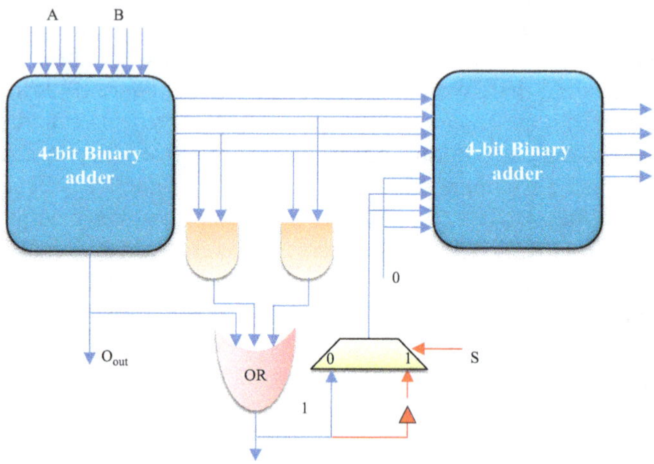

Figure 7.2 BCD adder (implemented using AND gate and OR gate) as a micro-3PIP as a component of the HLS tool library

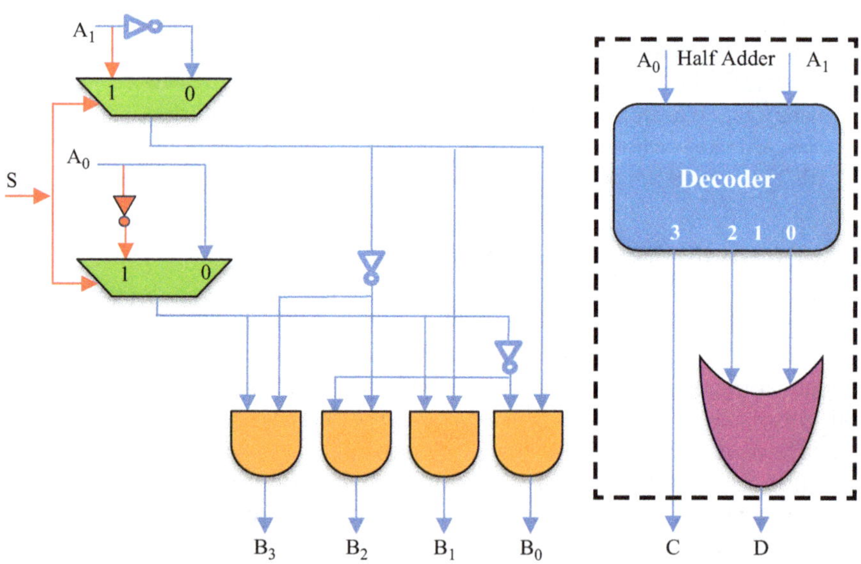

Figure 7.3 Half adder (implemented using a decoder and an OR gate) as a micro-3PIP as a component of the HLS tool library

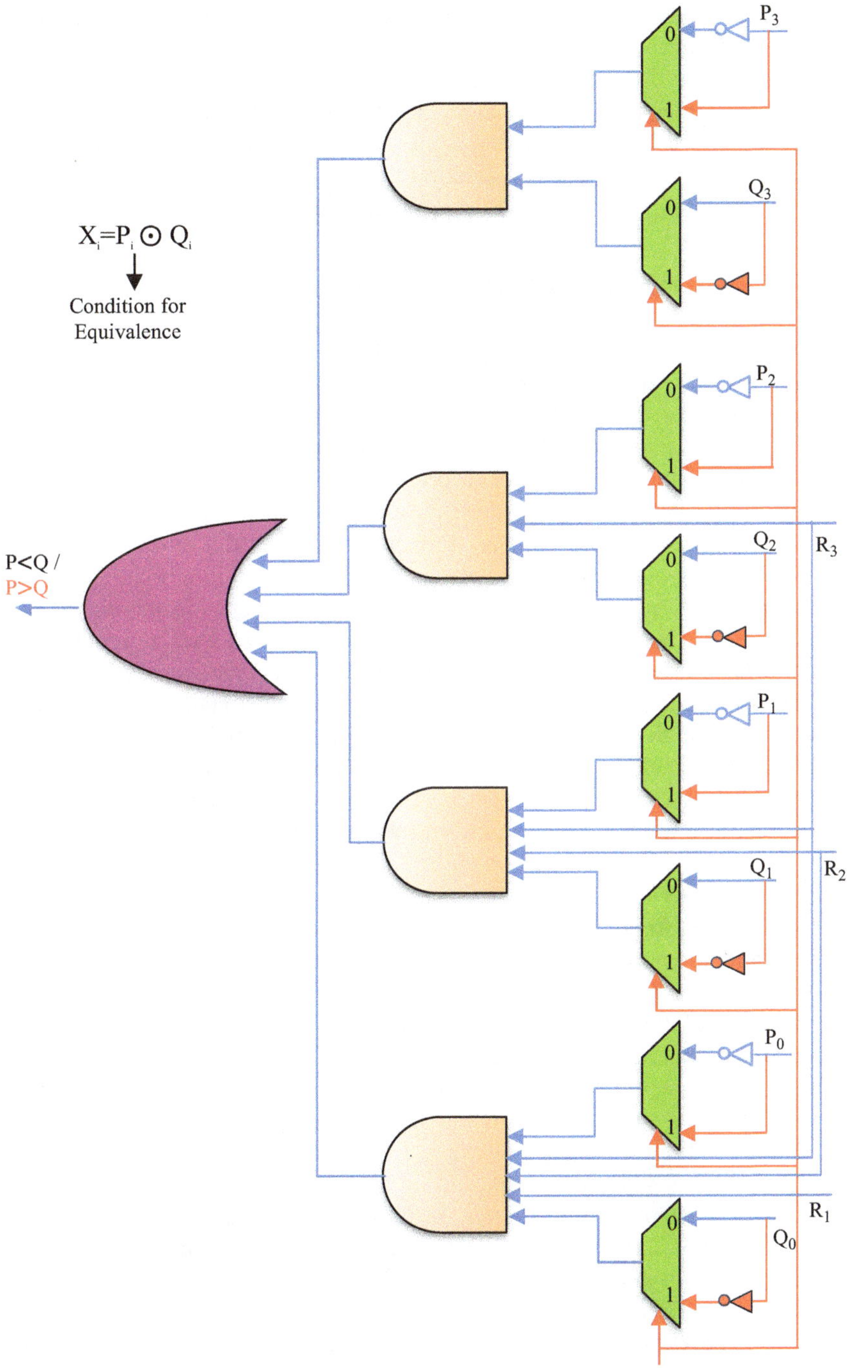

Figure 7.4 Comparator (implemented using an AND gate and an OR gate) as a micro-3PIP as a component of the HLS tool library

HLS tool library, which can easily compromise the HLS-designed hardware IPs (before they are integrated into the SoCs).

By narrowing the focus to the untrustworthiness of the third-party HLS tool library, this chapter aims to highlight the vulnerabilities introduced by external micro-3PIPs of the HLS tool library while assuming that the internal HLS design process is secure. Consequently, detection and mitigation efforts are concentrated upon using a distinct IP vendor allocation policy and DMR logic.

7.4 What is an HLS functional Trojan?

This section illustrates the hardware Trojan threat in 3PIP modules and highlights the importance of ensuring hardware security during the HLS process. In this context, an untrusted HLS tool developer may deliberately insert Trojan logic within a module or micro-3PIP component as part of the HLS tool library. The inserted Trojan logic typically remains dormant under normal operating conditions, only becoming active when triggered by a specific event or external condition. Because of this stealthy behavior, the Trojan is extremely difficult to detect during the in-house design team's standard validation or functional verification process.

Under normal conditions, the infected module appears to function correctly, producing valid output and conforming to the expected design specifications. Untrustworthy HLS tool providers generally supply register transfer level (RTL) files of these library modules, which may unknowingly contain such malicious or Trojan-infested logic. This chapter discusses the issue of Trojans that alter the computational behavior or output value of these modules once activated. However, a critical challenge arises because such Trojans manifest their malicious behavior only during runtime, after being triggered by an adversary. Prior to activation, they remain completely inactive and therefore undetectable through conventional RTL simulation or lower-level testing methods.

Even in cases where the Trojan is permanently active ("always-on"), its impact may be subtle, affecting only in a few cases while leaving the majority unaffected, thus escaping notice during routine simulation or verification. This challenge persists regardless of module size, as both small and large components within the HLS library can be exploited by adversaries in real-time operational environments to induce computational errors. Furthermore, even an exhaustive verification process, such as RTL-level simulation, physical inspection, or side-channel analysis may often fail to expose such threats, as exhaustive RTL functional validation of the IP designed is tedious and time-consuming.

The following section elaborates on the underlying reasons that make these types of hardware Trojans within IP modules so difficult to detect through conventional examination methods. Figures 7.2–7.4 illustrate a representative example of a Trojan embedded within a 3PIP or module in an HLS tool library.

In this context, several examples of potential hardware Trojan logic embedded within standard components of an HLS tool library, such as BCD adders, half adders, digital comparators, and multipliers, are considered. When triggered, these malicious circuits alter the normal functional output of the corresponding component. The activation or triggering of such hardware Trojans can occur through various mechanisms, including a Trojan time bomb based on a finite-state machine (FSM) counter, an external antenna, or some activation mechanism. Irrespective of the component's application within the overall IP design, once activated, the Trojan logic inevitably impacts the computational output value, thereby compromising the IP design functionality.

The particular class of Trojans explained in this chapter is designed specifically to manipulate the computational output values of the system, causing functional deviation rather than performing information theft or data leakage. Consequently, the attacker does not require prior knowledge of the system architecture to exploit these Trojans. Their sole purpose is to cause erroneous computation results upon Trojan activation. Detection techniques that operate at lower levels of abstraction, such as side-channel analysis (Wang *et al.*, 2008) and RTL simulation, often fail to expose these Trojans for several reasons. First, Trojan circuits are typically designed to activate only under very rare or specific predefined conditions (for instance, after a certain FSM counter value or under the influence of an external antenna). This selective activation behavior makes them highly unlikely to be triggered during standard functional verification using simulation test vectors (Tehranipoor and Koushanfar, 2010). Second, during RTL-level functional verification, the overall system design is typically validated as a whole, and the infected IP behaves as a legitimate component, thereby passing the verification process successfully. The Trojan remains dormant throughout these tests and only activates during real-time deployment under adversarial conditions to cause a functional anomaly. Third, in the case of 3PIPs, the absence of a trusted golden reference model makes Trojan detection practically infeasible during HLS. Moreover, techniques such as physical inspection and reverse engineering are not only prohibitively expensive but also highly complex due to the miniaturization of IC features at nanometer scales. Since Trojans are strategically inserted within specific regions of the design and may be activated by a single trigger signal, reverse engineering cannot guarantee detection. The accuracy and resolution limits of reverse engineering methods further complicate this process. Additionally, traditional manufacturing fault detection tests, including stuck-at fault, delay fault, and bridging fault analyses, are ineffective against such malicious logic. Similarly, detection through parametric signal analysis, such as monitoring power, delay, or frequency variations, has become increasingly unreliable with the ongoing reduction in transistor feature size and improvements in process technology (Tehranipoor and Koushanfar, 2010).

In the broader design flow, the HLS process converts the behavioral description of a design, represented as a CDFG, into an equivalent RTL implementation. This conversion inherently involves DSE, which evaluates multiple candidate design solutions based on key optimization objectives such as area, delay, and power consumption (Sengupta and Bhadauria, 2014; Haubelt *et al.*, 2008; Sengupta and Sedaghat, 2011; Krishnan and Katkoori, 2006; Ferrandi *et al.*, 2008; Zeng *et al.*, 2010; Mandal *et al.*, 2000; Ram *et al.*, 2012; Liu *et al.*, 2002; Deb and

Kalyanmoy, 2001; Das *et al.*, 2009). DSE encompasses several interdependent sub-processes, including scheduling, allocation, and binding (Krishnan and Katkoori, 2006). The complexity of DSE increases further with the inclusion of auxiliary design parameters such as the loop unrolling factor in CDFG-based applications, as this introduces an additional exploration dimension that must balance conflicting design constraints between area and performance. Therefore, when dealing with trusted hardware, the consideration of these parameters during DSE becomes an intricate and computationally demanding task, as elaborated in Section I.

7.5 Discussion of different Trojan detection techniques

An effort has been made to develop or design a low-cost Trojan detection method within the HLS tool library that optimizes both scheduling and loop unrolling, tailored to designer-defined security constraints (Sengupta *et al.*, 2017). At the same time, some techniques exist that primarily rely on code-coverage analysis and property checkers, which verify whether an IP satisfies specific properties. In the case of code-coverage analysis (Hu, 1997), a list of suspicious/distrustful signals is compiled from the RTL design. These signals are flagged as suspicious if they remain sturdy during the code-coverage analysis phase. The underlying assumption is that Trojans typically do not change their states unless they are triggered. Based on this, the state-of-the-art discusses additional test vectors to stimulate parts of the design that have not been covered by previous tests. While this technique is helpful, it significantly increases the verification time. Furthermore, redundant logic removal is employed as a part of this process, since certain logic tends to stay in a constant logical state during verification. This step involves inserting scan chains into the post-synthesis gate-level netlist. Analysis of stuck-at faults using test patterns generated by an automatic test pattern generation (ATPG) process is performed. A stuck-at fault that cannot be tested is often an indication of redundant logic. Therefore, the ATPG detects a stuck-at-1 or stuck-at-0 fault as untestable. The corresponding fault net can be superseded with a logical 1 or 0 in the gate-level netlist, and all circuits driving these fault nets are removed. Equivalence analysis (Bushnell and Agrawal, 2013), which is typically used for fault detection, can also be applied to the list of suspicious signals in an attempt to reduce the number of flagged signals. However, this introduces additional runtime overhead. Moreover, not all of the signals identified as suspicious are necessarily Trojans. The complexity of the Trojan inserted often causes the list of suspicious signals to grow as well. One of the significant challenges of this is the difficulty of achieving 100% code-coverage for all IPs, making it a less reliable solution in the long run.

Salmani *et al.* (2009) discussed an effective procedure for inserting a dummy flip-flop, which is employed in order to enhance the transition probability of nets when it falls below a predefined threshold. Transition probability means the likelihood that a signal changes state, directly impacting circuit performance and testability. To accurately evaluate this, the transition behavior is modeled using a geometric distribution, which estimates the probability of signal state changes over time. In addition to this, several methods have been discussed for detecting Trojans at lower levels of IP design (Tehranipoor and Koushanfar, 2010; Agrawal *et al.*, 2007; Narasimhan *et al.*, 2010; Jin and Makris, 2008; Wang *et al.*, 2008). However, apart from these state-of-the-art in Ram *et al.* (2012), which focuses on detection of Trojans at the system level, no effort has been made to explore the optimization of Trojan-secure scheduling combined with optimal loop unrolling (according to designer-defined constraints). Wang *et al.* (2008) highlighted a Trojan detection approach that involves generating a methodology based on integrating multiple supply transient currents. After a Trojan is detected, an isolation process is triggered to isolate the respective Trojan. Tehranipoor and Koushanfar (2010) discussed a detection process that relies on comparing the actual chip characteristics, such as path delay and power consumption, with expected values to identify any discrepancies that may validate the presence of a hardware Trojan. Additionally, the method in Narasimhan *et al.* (2010) is designed to detect malicious modifications in hardware, even in the presence of significant noise caused by process variations. The detection mechanism exploits the fact that Trojans, once inserted, alter the design parameters of the circuit, which can be observed through careful analysis of these changes.

Rajendran *et al.* (2013) and Cui *et al.* (2014) both have discussed Trojan detection at the system level, employing a concurrent error detection methodology for identifying hardware Trojans. These state-of-the-art leverages a diverse set of 3PIP vendors as part of the Trojan identification process. However, both studies primarily focus on DFGs and do not extend their techniques to handle CDFGs that involve loops, which limits their applicability in more complex designs where low-cost loop structures are prevalent during HLS. Additionally, these state-of-the-art lack methods for efficiently exploring IP vendor allocation strategies within hardware redundancy schemes such as DMR. Efficient vendor allocation is critical because it directly impacts the final area and delay characteristics of the implemented solution (Sengupta and Bhadauria, 2015). Without optimizing this allocation, the resulting hardware IP design tends to be less efficient, leading to higher costs and suboptimal performance. Moreover, Sengupta *et al.* (2017) focus exclusively on detecting hardware Trojans after their presence in the system, without addressing strategies to prevent hardware Trojan activation or mitigate their effects (Rajendran *et al.*, 2013). In contrast, the low-cost Trojan-secured HLS framework (Sengupta *et al.*, 2017) differs significantly from fault-secured HLS methods such as that described in Sengupta and Sedaghat (2015). These differences stem from variations in the threat model, the number of vendors required, and the nature of the security constraints imposed. Specifically, the low-cost hardware Trojan-secured HLS (Sengupta *et al.*, 2017) threat model assumes the presence of malicious logic inserted covertly by an adversary/attacker within the HLS tool library containing micro-3PIP modules/components, which remains dormant until triggered, representing a more insidious and targeted threat. On the other hand, (Sengupta and Sedaghat, 2015) fault-secured HLS is designed to provide resilience against transient faults, which are typically temporary in nature. Also, it is a self-inflicted malfunction caused by environmental factors rather than permanent malicious intent. These faults, commonly referred to as single-event upsets (SEUs), temporarily flip data without causing any permanent damage (Sengupta and Sedaghat, 2015).

Furthermore, (Sengupta *et al.*, 2017) hardware Trojan-secured HLS mandates the involvement of at least two distinct micro-3PIP vendors to ensure diversity and provide reliable Trojan detection in the DMR output. In contrast, (Sengupta and Sedaghat, 2015) fault-secured HLS requires only a single vendor, since the focus is on fault detection rather than functional Trojan detection. Additionally, the hardware Trojan-secured HLS framework enforces a strict requirement that sister operations in the DMR be allocated to hardware from different vendors to maximize security through diversity. Overall, these distinctions highlight the unique challenges and security considerations involved in designing a hardware Trojan-secured HLS framework compared to traditional fault-secured approaches.

7.6 Need for security-centric HLS

The digital integrated circuit undergoes design exploration at various levels of abstraction, in line with the "divide and conquer" philosophy. This requires managing the growing complexity of modern 3PIP design while simultaneously controlling the non-recurrent design cost. At the core of this process lies HLS, which has evolved significantly over time. Initially focusing on performance optimization, HLS has expanded to address other critical concerns, including power optimization and process variation management, all at the architectural level (Sengupta *et al.*, 2017; Mohanty *et al.*, 2002, 2014).

In today's age of advanced mobile computing, the demand for IP-based designs has skyrocketed to meet the pressing need for faster time-to-market delivery. HLS techniques play a crucial role in enabling design engineers to efficiently create such IP designs. This shift toward IP reuse, driven by HLS, reduces design time and cost, allowing engineers to focus more on higher-level design decisions and optimizations. Given this context, it is no surprise that one of the natural extensions of HLS research is to incorporate security awareness alongside other architectural challenges. As IC designs become more complex and more reliant on 3PIP, security has become an increasingly important consideration. While HLS has already proven its effectiveness in optimizing performance, power, and other factors, there was a need to extend these capabilities to address security threats, particularly those posed by malicious modifications, such as hardware Trojans. The inclusion of Trojan insertion risks in the HLS design flow introduces new challenges, as malicious micro-3PIPs can be covertly inserted into a design without being detected through traditional verification methods. This underscores the necessity of integrating security features into the HLS flow at the architecture level, where design decisions can have far-reaching implications for both performance and security.

In response to this growing concern, the research community has started to explore low-cost, Trojan-aware HLS solutions that can safeguard against these threats without compromising the benefits of HLS. One such state-of-the-art focuses on loop-based CDFGs, which are central to many high-level designs. Figure 7.5 illustrates a security-aware HLS solution specifically designed for this context, where the goal is to integrate Trojan detection and mitigation mechanisms at a low cost

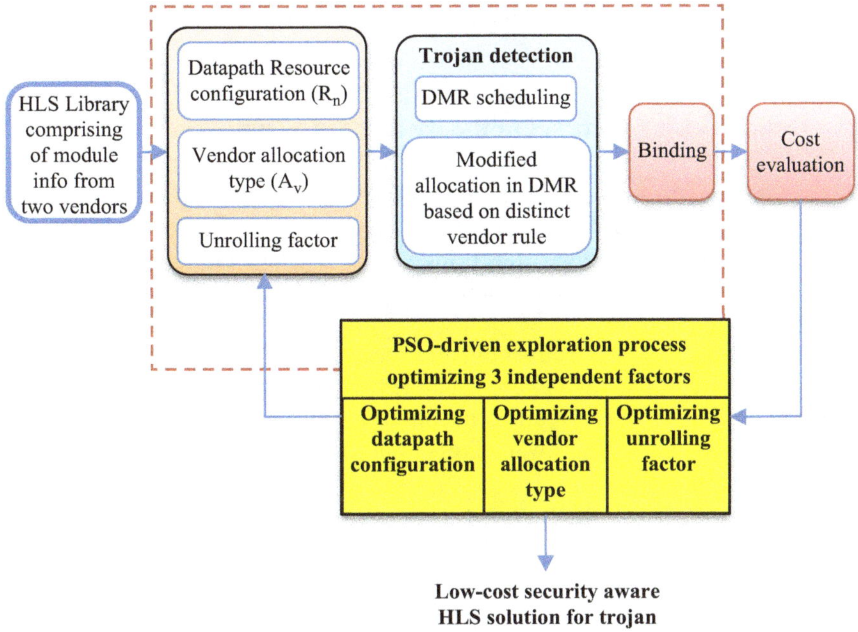

Figure 7.5 Low-cost Trojan-secured DMR schedule during HLS

while maintaining the overall efficiency of the design process. This figure includes a comprehensive overview of a low-cost, security-aware HLS designed for a Trojan-secured datapath IP. The process initiates with an HLS library comprising potentially untrustworthy micro-3PIP modules. Additionally, the technique requires similar micro-3PIP modules from two distinct IP vendors, thereby providing diverse design components. Key parameters include datapath resource configuration, vendor allocation type, and unrolling factor, which collectively influence the resource and scheduling decisions. Central to the framework is the Trojan detection block, which incorporates DMR scheduling and a modified allocation strategy based on distinct vendor rules to enhance security against hardware Trojans. The output from this block proceeds into the binding and cost evaluation stages, ensuring that the design not only meets security requirements but also remains cost-effective. A crucial feature is the PSO-driven exploration process, which simultaneously optimizes the datapath configuration, distinct vendor allocation, and unrolling factor, creating an iterative feedback loop that refines design parameters for optimal performance and security. This optimization ensures that the final solution balances the trade-off between design cost and security effectively.

The key challenge in this area is to ensure that security considerations do not introduce excessive overhead in terms of design complexity or cost. By embedding security features directly into the HLS process, it becomes possible to detect and neutralize potential Trojans before they can cause harm. This proactive approach helps designers detect security vulnerabilities early in the design process, reducing the risk of IP design malfunction.

7.7 Details of Trojan security-aware HLS process

This section aims to explain the design space exploration problem of a Trojan-secured DMR schedule. It focuses on candidate solutions, including DMR schedule resource configurations (architecture), loop unrolling factors, and distinct vendor assignment procedures. These components are hybrid-encoded, a method detailed in the following sections. The section will examine how different configurations and factors influence the effectiveness and security of the DMR schedule, providing insights into optimal choices for each element. By analyzing these factors, the section aims to identify robust, efficient solutions for securing the DMR scheduling process against functional hardware Trojan-based attacks.

7.7.1 Summary of the problem

The goal is to determine a solution for a Trojan-secured DMR schedule, optimizing the set $\{X_i\} = \{N(R_1), N(R_2), ..., N(R_D), UF, A_v\}$, while exploring the design space of a given CDFG (Control Data Flow Graph) and satisfying conflicting user constraints, ultimately minimizing the overall cost. The problem can be highlighted as follows:

Minimize [Hybrid Cost ($A^{DMR}{}_T$, $T^{DMR}{}_E$) for Optimal $\{X_i\}$], subject to the constraints: $A^{DMR}{}_T \leq A_{cons}$, $T^{DMR}{}_E \leq T_{cons}$, and hardware Trojan security.

Here, X_i is the resource set of a particular solution with unrolling and allocation information. $N(R_1)$, $N(R_2)$, ..., $N(R_D)$ is the number of instances of resource type "1," . . . . . ,"d," $A^{DMR}{}_T$ is the total area of a DMR design, $T^{DMR}{}_E$ is the total execution time of a DMR design, A_{cons} is the user-specified area constraint, T_{cons} is the user-specified execution time constraint. The vendor allocation procedure, denoted by "A_v," can only hold a binary value. When "A_v" = 1, all operations within a specific unit are assigned to resources from the same vendor type (e.g., all operations of the original unit are assigned to vendor "V_1," and all duplication operations to vendor "V_2"). When $A_v = 0$, alternate vendor assignments are made for operations within a unit's CS (Critical Section). The A_v variable plays a critical role in optimizing the Trojan-secured schedule, as both $A_v = 0$ and $A_v = 1$ ensure vendor distinctness in the DMR design, enhancing Trojan security. However, these two values impact the final delay of the design differently.

7.7.2 Models used for design cost evaluation

In the state of the art, each particle's position represents a hybrid encoding of candidate solutions within the design space. This includes schedule resource configurations (X_i), the loop unrolling factor (UF), and the vendor assignment procedure (A_v). These components are combined to explore potential solutions, with each particle in the search space corresponding to a unique combination of these parameters. This allows for efficient optimization of the DMR schedule by evaluating various configurations and their impact on performance, cost, and security.

i) Area model: The total area consumed ($A^{DMR}{}_{T}$) by a resource set is calculated as follows:

$$A_T^{DMR} = \sum_{j=1}^{2} \sum_{i=1}^{m} \left(P\left(S_i^{V_j}\right) X S_i^{V_j} \right) \tag{7.1}$$

where, S_i^{Vj} is the number of instances utilized from vendor V_j for a resource type S_i, and "m" is the maximum number of instances of resource type S_i for vendor V_j, while $P(S_i^{Vj})$ is the area of a resource type (S_i) corresponding to vendor (V_j).

ii) Delay model:

Case 1: When "U" is equal to one (indicates no unrolling), then the total # of control steps (CS) = # of CS required for executing the loop body once * number of duplicate iterations of the loop body:

$$C_T{}^{DMR} = C_{body}^{DMR} * \beta = C_{body}^{DMR} * [1/\mathrm{UF}]^{\mathrm{floor}} \tag{7.2}$$

where β= $[I/UF]$ ^{floor}

$$\text{Since UF} = 1,\ C_T{}^{DMR} = C_{body}^{DMR} * \mathrm{I} \tag{7.3}$$

where $C_T{}^{DMR}$ is the total CS required to execute the loop of CDFG DMR completely, $C^{DMR}{}_{body}$ is the number of CS required to execute the loop body of CDFG DMR once, and "I" is the maximum number of iterations (loop count).

Case 2: When "*UF*" unevenly divides I: In such a case, *I mod UF* iterations will be executed sequentially; therefore, the total number of CSs is:

$$C_T{}^{DMR} = \left(\mathrm{C}_{\mathrm{body}}^{\mathrm{DMR}} * [1/\mathrm{UF}]^{\mathrm{floor}}\right) + (ImodUF) * \mathrm{C}_{\mathrm{first}}^{\mathrm{DMR}} \tag{7.4}$$

where $C^{DMR}{}_{first}$ is the number of CS required to execute the first iteration of the $CDFG^{\ DMR}$. Hence, the execution time for the system is calculated as:

$$T_E{}^{DMR} = \theta * C_T{}^{DMR} \tag{7.5}$$

where "Θ" is the delay of one CS in nanoseconds.

Fitness function: The fitness function (considering execution time and area consumption of a solution) is inspired by Kumar *et al.* (2010) and is formulated as:

$$C_f(X_i) = W_1 \frac{A_T{}^{DMR} - A_{cons}}{A_{\max}{}^{DMR}} + W_2 \frac{T_E{}^{DMR} - T_{cons}}{T_{\max}{}^{DMR}} \tag{7.6}$$

where $C_f(X_i)$ is the cost of the particle with resource set X_i, $A_{max}{}^{DMR}$ and $T_{max}{}^{DMR}$ are the DMR solutions with maximum area and delay in the design space, and W_1 and W_2 are the user-defined weights, both kept at ½ during exploration to provide equal preference.

iii) PSO encoding and upgradation: The particle is encoded as follows:

$$X_i = (M(R_1)(M(R_2),..(M(R_d)..(M(R_D), UF, A_V) \quad (7.7)$$

Here, $M(R_D)$ represents the number of resources of type D, UF indicates the unrolling factor, and A_v denotes the distinct vendor allocation type (encoded as 0 or 1).

In PSO-DSE, each dimension (*d*) of a particle position X_i (except the last dimension) is updated using the following function (Crop *et al.*, 2010):

$$R_{d_i}^{+} = R_{d_i} + V_{d_i}^{+} \quad (7.8)$$

The variable $V_{d_i}^{+}$ is updated by (9) as follows:

$$V_{d_i}^{+} = \omega V_{d_i} + b_1 r_1 [R_{d_{lbi}} - R_{d_i}] + b_2 r_2 [R_{d_{gb}} - R_{d_i}] \quad (7.9)$$

where, $R_{d_i}^{+}$= new resource value or "U" value of particle X_i in d^{th} dimension and R_{d_i} = resource value or "U" value of particle X_i in d^{th} dimension. $V_{d_i}^{+}$ = new velocity of ith particle in d^{th} dimension, $R_{d_{lbi}}$ = resource value of X_{lbi} (local best position) in d^{th} dimension, ω = inertia weight, b_1, b_2 = acceleration coefficients. The process of local and global best upgradation is explained in Mishra and Sengupta (2014).

7.8 PSO-driven exploration of Trojan-secured DMR datapath

The DSE process, driven by PSO, integrates multiple inputs to efficiently explore hardware micro-3PIP design solutions for Trojan detection. Key inputs include a module library, the behavioral description of a CDFG, predefined user parametric constraints for area and delay, control parameters of the PSO (such as inertia weight, acceleration coefficients, and swarm size "p"), a maximum iteration count for the CDFG, and the preprocessing algorithm for unrolling factors. These inputs serve as the foundation for the DSE process, as illustrated in Figure 7.6. A brief overview of the process is provided below for immediate reference.

The PSO-based DSE process begins with an innovative particle encoding scheme that represents a candidate/designer design solution, labeled as X_i, within the design space. This encoding includes three key components: the DMR schedule architecture (denoted as $\vec{R}_n$), the unrolling factor (denoted as UF), and the vendor allocation procedure type (denoted as A_v). In the PSO-DSE framework, a particle representing a potential design solution is designated as X_i, where "I" identifies its position within the population of candidate solutions.

$$X_i = \left(\vec{R}_n, \mathrm{UF}, \mathrm{Av}\right) \quad (7.10)$$

The $\vec{R}_n$ represents the resource configuration, such as the number of adders, multipliers, and other functional units required for the scheduling. The inclusion of the A_v dimension, indicating the distinct vendor allocation procedure, is crucial because it reflects the distinct allocation of vendor types, which directly affects the design's

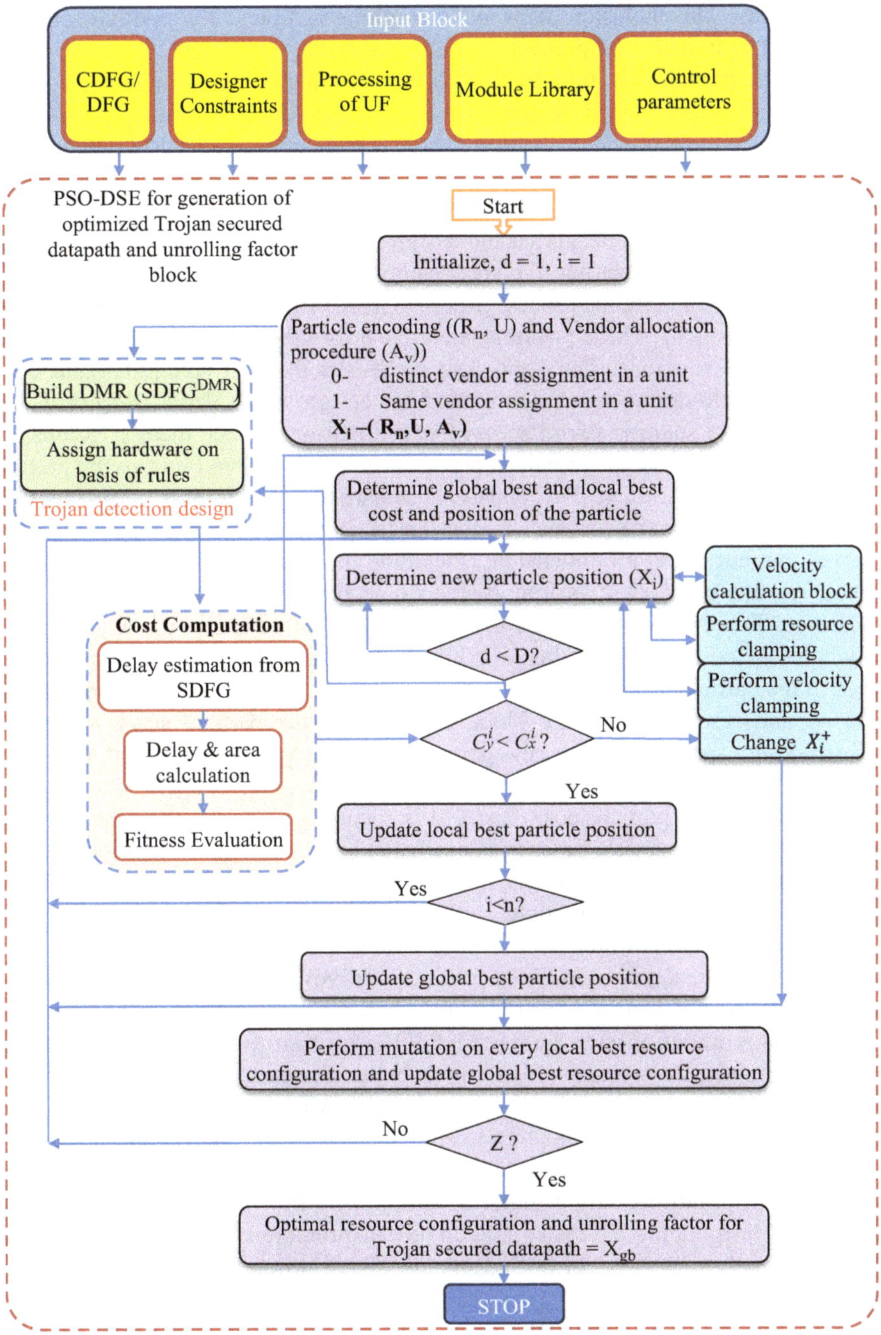

Figure 7.6 Trojan security-aware PSO-DSE methodology

area and delay. This inclusion helps avoid overlooking potentially optimal solutions during the Trojan detection process, as discussed in the earlier section.

Once the particles are initialized with the encoding scheme, the next step is to build a Trojan-secured DMR design for the given candidate solution X_i. This design is then subjected to a fitness evaluation, which is computed using a cost computation block. The cost computation involves calculating the area and delay metrics of the system to assess the efficiency of the design. The fitness value associated with each particle reflects how well it satisfies the design constraints, including minimizing the area and delay while ensuring Trojan detection and security. After evaluating the initial candidate solution, the process proceeds to determine the local best and global best particle positions. These positions are crucial for guiding the swarm of particles toward optimal solutions. The local best represents the best solution found by a particular particle during its exploration of the design space, while the global best refers to the best solution found across the entire swarm. Once these positions are determined, the next step is to calculate the particle velocity and update the particle position (X_i^+) accordingly. The update of a particle's position is carried out by evaluating the new value for each dimension (d) in the particle's position, X_i. This process continues until all dimensions, including the UF and vendor allocation procedure A_v, are checked. It is important to note that in this framework, the final dimension "D" corresponds to the UF, as indicated in (10), because the A_v dimension only holds a Boolean value (either 0 or 1), which represents the allocation procedure types. If the particle's position violates any boundary conditions, interaction with the velocity and resource clamping block ensures that the particle stays within the defined design space. Once the new particle position (X_i^+) is determined, a Trojan-secured DMR design is again built for this new position, and its cost is computed. If the cost of the new particle position (X_i^+) is found to be higher than the previous position (X_i), a new particle position is evaluated to find a better solution. If the cost is lower, the local best position is updated with the new solution. This iterative process continues for each particle in the swarm population, and the global best position is updated accordingly.

Mutation: The next phase of the process involves a mutation step, which aims to further refine the design by introducing small random changes to the particle's position. This mutation step helps the PSO algorithm avoid local minima and explore a broader design space, leading to better design solutions. The mutation process encourages diversity within the swarm, which can help discover new, improved design configurations that would not have been found through standard velocity updates alone. The mutation operation is applied to all local best resource configurations with a mutation probability $P_m = 0.25$. This operation is crucial for introducing diversity and avoiding premature convergence by perturbing the solutions in a controlled manner. Two primary mutation techniques are utilized:

1) the rotation operation, and
2) the increment or decrement operation.

To effectively apply these mutations, the total population of candidate solutions is divided into two distinct groups based on their index:

Even Group: For particles in the even group, the algorithm applies a left rotation operation. This involves shifting the elements of the resource configuration by one position to the left, wrapping the elements around in a circular manner. This operation alters the configuration while maintaining the overall structure of the solution, potentially leading to a new, viable solution.

Odd Group: For particles in the odd group, the algorithm performs an increment or decrement operation. A random value is generated, and the resource configuration is either incremented or decremented by this random number. This operation introduces variability in the solution space, allowing the particle to explore different regions in search of better configurations. The combination of these mutation techniques ensures a balanced exploration of the search space and helps maintain diversity within the population of solutions.

The entire PSO-driven DSE process continues iteratively until the terminating criteria (Z) are met. These criteria may include a specified number of iterations or a convergence threshold that indicates when the algorithm has found an optimal solution or when further exploration would yield minimal improvements. The final output of the DSE process is the global best particle position, which corresponds to the optimal hardware IP design for Trojan detection, satisfying all area, delay, and security constraints. In summary, the PSO-based DSE framework is a powerful tool for optimizing hardware designs, particularly for systems requiring Trojan detection and security. By incorporating a variety of inputs, including resource configurations, unrolling factors, and vendor allocation procedures, the framework explores the design space efficiently and effectively. The iterative process, combined with fitness evaluation, velocity updates, and mutation, ensures that the final design is both secure and optimal in terms of area and delay. The PSO-driven approach ultimately enables the identification of an efficient and cost-effective Trojan-secured DMR design.

7.9 Designing a Trojan-secured DMR schedule for the design solution during DSE

To design a Trojan-secured schedule, the DMR logic incorporates a specific vendor allocation rule. The process begins by generating the corresponding DMR schedule. DMR logic duplicates all unrolled operations in the CDFG based on UF from the candidate/designer design solution X_i, as detailed in (10). This step ensures that DMR is applied to the CDFG at the behavioral level before any scheduling takes place. After unrolling the CDFG operations, both the original and duplicate operations are scheduled simultaneously. A list scheduling algorithm is used, relying on the schedule architecture details of the candidate design solution (Xi), such as [$N(R_1)$, $N(R_2)$, ..., $N(R_D)$]. This approach enables the extraction of information related to the $C^{DMR}{}_{first}$ and $C^{DMR}{}_{body}$. By doing so, the final delay, denoted as $C_T{}^{DMR}$, can be determined, as discussed in Section 7.2.2. This method ensures that the Trojan-secured schedule meets the required redundancy and delay specifications.

The DMR logic, combined with a specific vendor allocation rule, is employed to design a Trojan-secured schedule. The vendor allocation rule is a critical part of ensuring Trojan security in terms of Trojan detection. It states that for the DMR design, two distinct vendors must be assigned to similar operations in the original and duplicate units. Specifically, the operation (v) of the original unit (Z^{OG}) and the operation (v‘) of the duplicate unit (Z^{DP}) must be allocated to different micro-3PIP vendors. This rule is crucial because it eliminates the coincidence of a Trojan payload. It is extremely unlikely that Trojans in different third-party micro-3PIPs will produce the same incorrect output. This unique vendor allocation prevents potential Trojans from simultaneously corrupting both copies of the operation, making the system more secure. The primary focus is on ensuring Trojan security through distinct vendor assignments. It is important to note that the rule for distinct vendor allocation can be implemented in several ways. This section explains the most optimal allocation method for similar operations in both the original and duplicate units. This exploration aims to maximize both security and system efficiency. The allocation of distinct vendors within the DMR schedule significantly influences the overall system performance. Specifically, the assignment of vendor IPs to operations controls the final delay and area of the design. Different vendors may offer similar resources, but their implementation can differ in terms of area and delay. For example, two different vendors may provide similar resources, but one might have a larger area requirement or incur more delay than the other. This variation is crucial because the delay and area affect the system's overall performance.

Thus, the vendor allocation not only impacts Trojan security but also directly affects the design's efficiency. By carefully selecting how to distribute operations between vendors, the overall system can be optimized to achieve the best possible balance of security, delay, and area. In summary, the key to the hardware Trojan-secured schedule lies in the careful vendor allocation process. By ensuring that the original and duplicate operations are assigned to distinct vendors, the system's security is enhanced. At the same time, the choice of vendor influences the final delay and the area of the design. For determining the optimal vendor allocation, balancing both Trojan security and system performance to create the most efficient design solution.

In this section, we analyze specific IP characteristics from two distinct vendors, V_1 and V_2 (adopted from Sengupta *et al.*, 2017). The multiplier and adder provided by vendor V_1 have an area of 2468 and 2034 au, respectively. Their delays are 10,000 ns for the multiplier and 265 ns for the adder. Vendor V_2 provides a multiplier with an area of 2464 au and an adder with an area of 2032 au. Their delays are 11,000 ns and 270 ns, respectively. These parameter values are critical to the state-of-the-art (Sengupta *et al.*, 2017). They are considered during the design of an optimized Trojan-secured schedule. Since these micro-3PIP components are used as modules in the HLS tool library, any differences in area or delay across vendors directly affect the final solution. The design outcome changes depending on which vendor's components are used. Two different methods of vendor allocation within the DMR schedule are explored. These are labeled as $A_v = 1$ and $A_v = 0$.

7.9.1 Type 1: Vendor allocation process ($A_v = 1$)

In this method, all operations of one unit, either the original (Z^{OG}) or duplicate (Z^{DP}), are assigned to a single vendor. For example, all operations in Z^{OG} may use micro-3PIPs from vendor V_1, and all in Z^{DP} may use micro-3PIPs from a distinct vendor V_2. This ensures that every pair of similar operations between Z^{OG} and Z^{DP} uses different vendors.

7.9.2 Type 2: Vendor allocation process ($Av = 0$)

In this method, alternate vendor assignment is applied within a control step (CS) of a unit. That means operations using the same type of resource are alternately assigned to vendors V_1 and V_2. If a new type of resource appears within the CS, the alternate assignment starts independently for that type.

For example, as shown in Figure 7.8, in CS (1), operations 1, 2, 3, and 10 are assigned to V_1, V_2, V_1, and V_2 respectively. In another CS (4), for subtractors, operations 7 and 15 are alternately assigned to V_1 and V_2. Similarly, for multipliers, operations 22 and 21 are also alternately assigned.

In both $A_v = 1$ and $A_v = 0$, similar operations in Z^{OG} and Z^{DP} are always assigned to different vendors. This ensures the Trojan detection capability remains intact.

Now, consider two scenarios from the different benchmark, unrolled three times. In both cases (Z^{OG} and Z^{DP}), the resource constraints are: two adders (+), four multipliers (*), two subtractors (−), and one comparator (<), with UF = 3 and iteration count (I) = 4.

For $A_v = 0$, the resulting execution time (${T^{DMR}}_E$) is 176,350 ns, and the design area (${A^{DMR}}_T$) is 28,380 au.

For $A_v = 1$, the execution time is 173,350 ns, and the area is 38,908 au.

These results show a trade-off: $A_v = 1$ offers a slightly faster schedule but uses more design area, while $A_v = 0$ uses less area but takes more execution time.

It is observed that the two generated scheduling solutions have differences in both area and delay values. This difference arises even though both schedules use distinct vendor assignments for similar operations to ensure Trojan detectability. The DMR schedules shown in Figures 7.7 and 7.8 follow the same principle; every operation in the original unrolled CDFG is duplicated, and both versions are assigned to distinct vendors to provide Trojan detection security.

In both cases, two distinct vendors are used, and duplication of all operations is done to create a secure DMR schedule. For demonstration, consider a candidate/designer design solution:

$$X_i = (2(+),\ 4(*),\ 2(-),\ 2(<),\ UF = 3)$$

This indicates that the CDFG must be unrolled three times, producing iterations I1, I2, and I3. After unrolling, every operation across these three iterations is duplicated to form the unrolled DMR-CDFG.

Once the DMR version of the CDFG is created, both the original and duplicate operations are scheduled in parallel. The scheduling uses the available resource

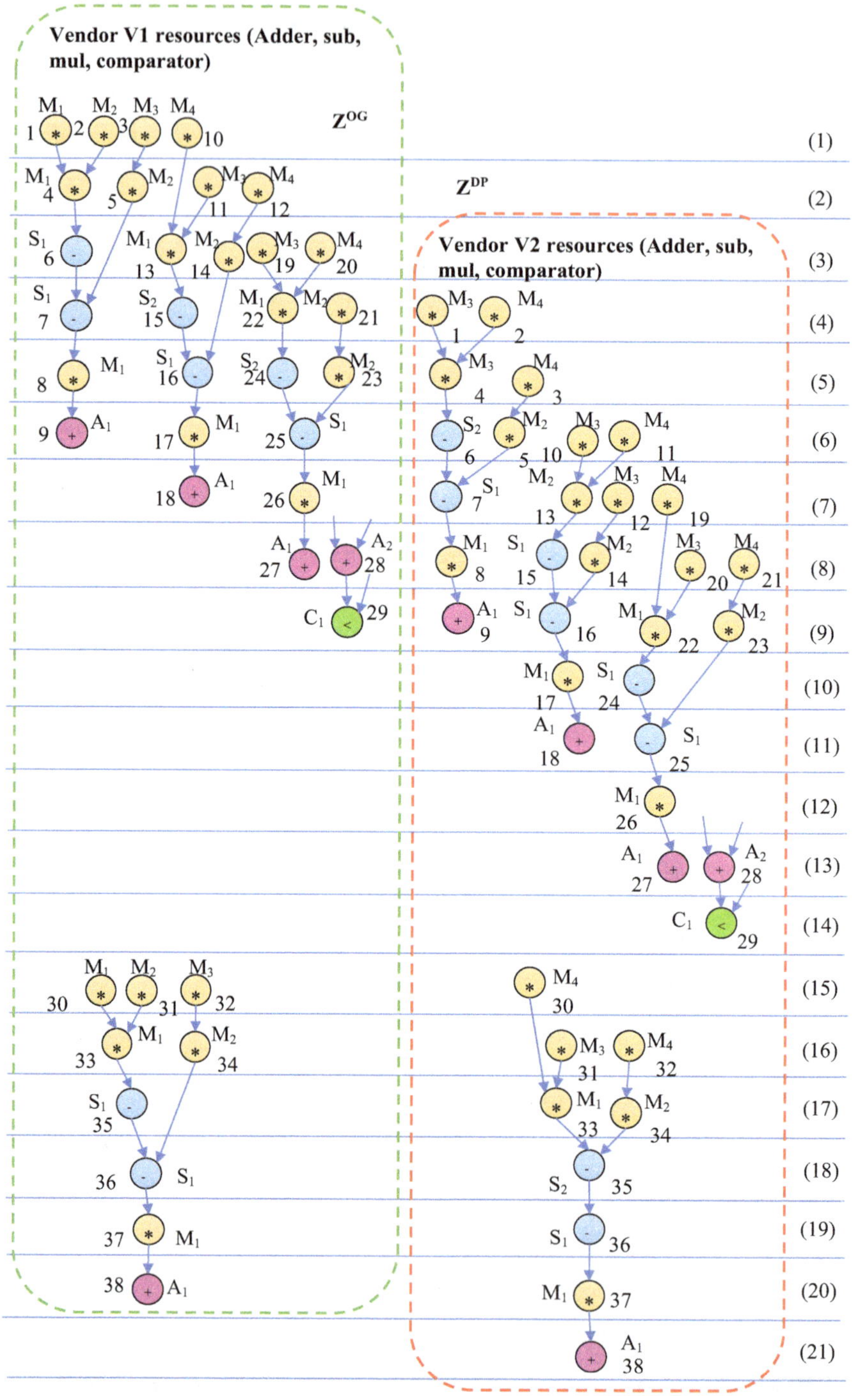

Figure 7.7 Scheduled and binded DMR design of differential equation for Xi = {(2(+), 4(*), 2(−), 2(<), U = 3, Av} with Av = 1.

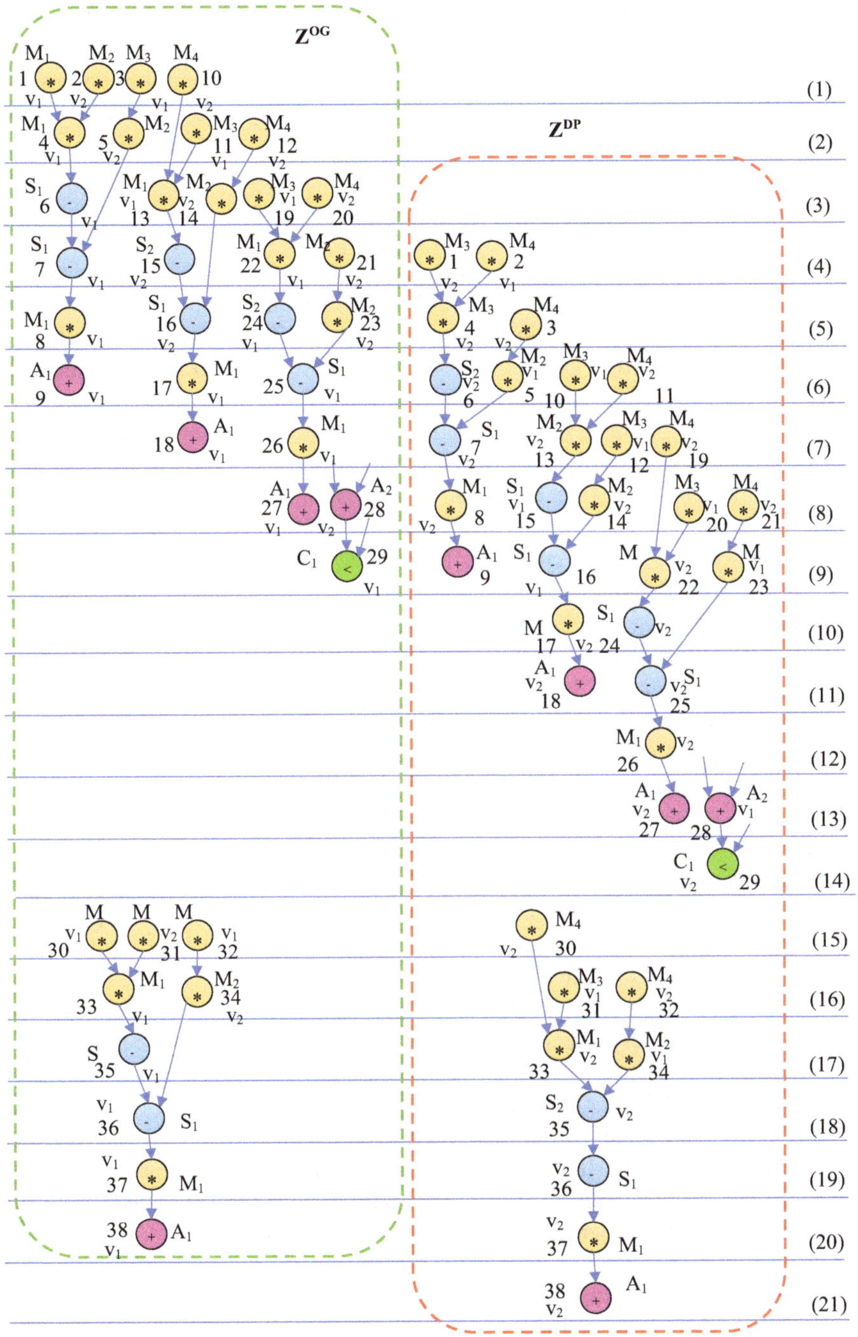

Figure 7.8 DMR design of differential equation for Xi = {(2(+), 4(*), 2(−), 2(<), U= 3), Av} with Av = 0

types mentioned in X_i: two adders (+), four multipliers (*), two subtractors (−), and two comparators (<).

However, despite having the same resource constraints and unrolling information, the two schedules shown in Figures 7.7 and 7.8 perform differently. One schedule may offer better delay, while the other may provide lower area usage. This variation depends solely on the value of A_v (either 0 or 1), which represents the vendor allocation method used. Although both ensure Trojan security by assigning similar operations from original and duplicate units to distinct vendors, they lead to different design metrics.

Therefore, when exploring DSE for Trojan-secured scheduling, the choice of A_v becomes crucial. Whether A_v is 0 or 1, it significantly impacts the final delay and area. Even though the schedule resources and unrolling factor remain the same, the vendor allocation strategy introduces notable differences in the outcome. Thus, during the DSE process, A_v must be considered an important design parameter. It adds an extra dimension to the exploration, alongside the resource array given in expression (10). By including A_v in the exploration, it is possible to identify the most optimal DMR schedule in terms of both area and delay, while still maintaining strong Trojan security.

The cost of the Trojan-secured DMR schedule is carefully evaluated in this case study (Sengupta *et al.*, 2017). The cost function takes into account several components. These include the area of functional resources, interconnect units like multiplexers (mux) and demultiplexers (demux), and comparators. Additionally, it includes the overhead caused by internal buffering, which is needed for temporarily storing operation outputs. Internal buffering is essential because similar operations in the original and duplicate units may complete at different times. To ensure proper comparison, the system stores outputs from both units in internal buffers until both are ready. This buffering adds extra area overhead to the design. The evaluation of design solutions, represented as particle positions in the search space, is done through a DSE process using PSO. This iterative process aims to find an optimal hardware Trojan fault-secured DMR system. The design must satisfy key constraints: area constraints (A_{cons}), timing constraints (T_{cons}), and minimize a combined hybrid cost that considers both.

From this discussion, three important factors become clear. First, efficient and distinct vendor allocation within the DMR scheduling is vital. Assigning operations from original and duplicate units to different vendors improves Trojan security. Second, the combination of scheduling resources is optimized. This helps balance the use of functional units and interconnects, reducing area and delay. Third, unwanted loop unrolling factors are screened out. This step avoids unnecessary overhead and helps achieve a low-cost, optimal Trojan-secured schedule. These three factors work together to enhance the overall design quality. Other state-of-the-art did not consider all these aspects together. Most earlier methods focused on either scheduling or security, but not both comprehensively. State-of-the-art fills that gap by jointly optimizing distinct vendor allocation, resource scheduling, and loop unrolling factors (Sengupta *et al.*, 2017).

By integrating these elements, the method achieves better results. It reduces area overhead, meets timing goals, and strengthens Trojan detection capabilities. The use of PSO helps explore a large design space efficiently. It guides the search toward the global best combination of parameters.

7.10 Terminating criteria

The PSO-DSE process integrated with Trojan security stops running under two conditions. First, it terminates if the number of iterations exceeds a set maximum limit, which is 100 in this case. This condition ensures that the algorithm does not run indefinitely and provides a hard stop after a reasonable amount of exploration. Second, the algorithm also terminates if there is no improvement in the global best solution over a certain number of consecutive iterations. Specifically, if the global best remains unchanged for ten iterations (denoted as (delta = 10)), the algorithm concludes. This condition helps avoid unnecessary computation when the solution has converged. Thus, the process terminates either when it reaches 100 iterations (condition S1) or when the global best solution remains unchanged for ten iterations without improvement (condition S2). Both conditions work together to balance optimal exploration at acceptable computational runtime.

7.11 Analysis of case studies

The technique was implemented in Java on an Intel Core i5-3210M CPU with 3 MB L3 cache, 4 GB DDR3 RAM, and a 2.5 GHz processor. An average of 15 runs was executed for small to large size problems (e.g., CDFG applications with over 200 nodes) while exploring optimal Trojan-secure solutions. Key PSO settings included: linearly decreasing inertia weight (0.9 to 0.1), acceleration coefficients $b_1 = b_2 = 2$, and stopping criteria = S1 or S2.

Figure 7.9 illustrates the impact of swarm size (p) on the exploration time of the PSO-DSE methodology. As seen, the exploration time increases with the swarm

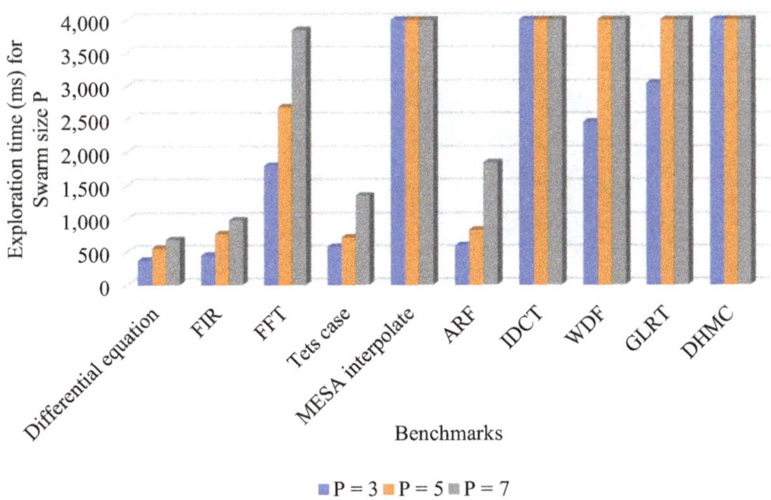

Figure 7.9 Comparison of exploration time with respect to swarm size p

size for all benchmarks, but this increase in time does not lead to an improvement in the quality of the Trojan-secure solutions found. This is because the total number of positions evaluated during each run rises with a larger swarm size, which in turn lengthens the exploration time. However, no significant gain in the solution quality is observed as a result of this increase in "p." For all the benchmarks, including single and nested loop DFGs, the PSO-DSE consistently generates solutions that adhere to the user-defined area-delay constraints. Additionally, it minimizes the hybrid cost described in (6), as demonstrated in Figure 7.10.

The cost function used in both the Trojan-secure framework (Sengupta *et al.*, 2017) and the method discussed in Rajendran *et al.* (2013) takes into account the area of the single comparator/error detection block responsible for runtime Trojan detection at the final output. Figure 7.11 provides a comparative analysis of the two state-of-the-art, assessing the design cost of the final architectural solution. Both methods leverage the concept of diverse 3PIP vendors, offering comparable security in terms of robustness. However, the PSO-DSE integrated with the Trojan-secure framework in Sengupta *et al.* (2017) consistently generates solutions with lower costs and, in most cases, outperforms (Rajendran *et al.*, 2013). The reasons for this improvement are outlined below:

Vendor Allocation Optimization: The PSO includes an additional encoding option that explores optimal distinct vendor allocation (A_v). This enhances the final optimization quality, resulting in a more cost-efficient solution.

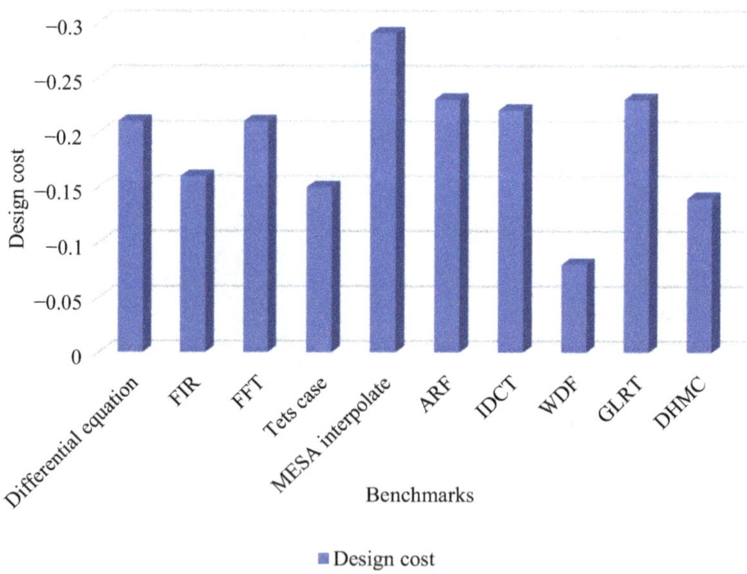

Figure 7.10 Design cost for different benchmarks

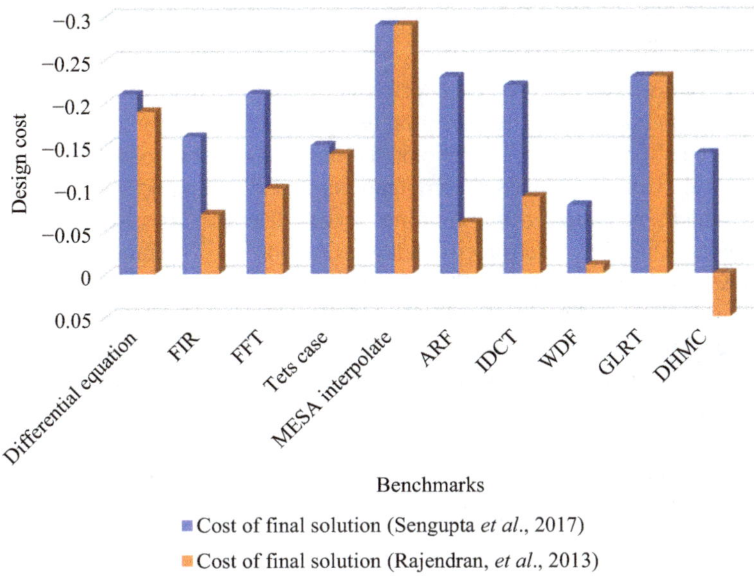

Figure 7.11 Design cost comparison between two state-of-the-art techniques

Unrolling Factor Optimization: Another encoding option in the PSO-DSE framework explores the optimal unrolling factor, improving the overall design efficiency and lowering the cost.

User-Constrained Resource Exploration: The PSO-DSE framework also explores scheduling resources within user-defined constraints, contributing further to cost reduction and better final design quality.

As a result, the case study of Sengupta *et al.* (2017) offers an average cost reduction of 11.4% for the final Trojan-secured schedule compared to Rajendran *et al.* (2013). Unlike the study of Sengupta *et al.* (2017), the study of Rajendran *et al.* (2013) does not have the ability to explore optimal unrolling factors or distinct vendor allocations, leading to higher costs. In Rajendran *et al.* (2013), the maximum unroll factor is used, and all units are allocated the same vendor ($A_v = 1$), resulting in higher overall costs. On the other hand, exploration of optimal distinct vendor allocation always leads to designing a cost-efficient final solution (Sengupta *et al.*, 2017).

7.12 Conclusion

This chapter focused on the simultaneous exploration of an optimized Trojan-secured schedule and an optimal loop unrolling factor for CDFGs. The PSO-DSE framework explains the challenge of generating hardware architectures that are not only performance-efficient but also secure against hardware Trojans. The chapter highlights several key features:

Execution Delay Modeling for DMR-Based Trojan Security: A comprehensive model was discussed to determine the execution delay of a DMR system designed to secure CDFGs against potential HLS functional Trojan attacks. This model accurately captures the behavior of redundant computations and their impact on overall performance when integrated with runtime error detection mechanisms, such as comparators. This allows for precise delay estimation while maintaining security.

Advanced particle encoding in PSO Framework: An advanced particle encoding scheme was discussed within the PSO-DSE framework, along with its integration with the Trojan-secure framework. This encoding supports simultaneous exploration of three critical design parameters: optimal schedule configuration, optimal loop unrolling factor, and optimal distinct vendor allocation (A_v). The inclusion of unrolling and vendor diversity into the particle structure allows the process to efficiently navigate the design space for solutions that are secure, cost-effective, and performance-optimized.

Overall, the chapter summarized the process of a low-cost optimal Trojan-secure framework that exploits PSO-DSE. The highlighted methodology is capable of handling a functional HLS Trojan.

7.13 Questions and exercises

1. What is micro-3PIP?
2. How micro-3PIPs are vulnerable to Trojan attacks?
3. What is a functional Trojan in HLS?
4. What are the threat models in hardware IP design?
5. What are some common techniques to detect hardware Trojans?
6. What is the Trojan security-aware HLS process?
7. What components are included in the particle encoding scheme used in the PSO framework?
8. How does PSO contribute to the optimization process?
9. What is the stopping criterion of the PSO process?
10. What is the significance of using DMR in securing CDFGs against hardware Trojans, and how does it affect execution delay?
11. How does the PSO-DSE differ from other state-of-the-art in terms of handling loop unrolling and vendor diversity during scheduling?
12. What role does the vendor allocation parameter (A_v) play in the final optimization process?
13. How does vendor allocation variation ($A_v = 0$ vs. $A_v = 1$) influence Trojan-secure design cost?
14. What are the advantages of simultaneously exploring loop unrolling, scheduling, and vendor assignment in a single optimization pass rather than separately?

References

A. Sengupta, A. Anshul, V. Chourasia, and N. Bhui, "Security vulnerability (backdoor Trojan) during machine learning accelerator design phases," *IEEE IT Professional*, vol. 27, no. 1, pp. 65–72, 2025.

A. Sengupta, A. Anshul, V. Chourasia, and N. Kumar, "M-HLS: malevolent high-level synthesis for watermarked hardware IPs," *IEEE Embedded Systems Letters*, vol. 16, no. 4, pp. 497–500, 2024.

A. Sengupta, S. Bhadauria, and S. P. Mohanty, "TL-HLS: methodology for low cost hardware Trojan security aware scheduling with optimal loop unrolling factor during high level synthesis," *IEEE Transactions on Computer-Aided Design of Integrated Circuits and Systems*, vol. 36, no.4, pp. 655–668, 2017.

V. K. Mishra, and A. Sengupta, "MO-PSE: adaptive multi objective particle swarm optimization based design space exploration in architectural synthesis for application specific processor design," *Elsevier Journal on Advances in Engineering Software*, vol. 67, p. 111124, 2014.

X. Zhang, and M. Tehranipoor, "Case study: detecting hardware trojans in third-party digital IP cores," *Proc. IEEE Int. Symp. Hardw. Oriented Security Trust (HOST)*, San Diego, CA, USA, pp. 67–70, June 2011.

M. Banga, and M. Hsiao, "A novel sustained vector technique for the detection of hardware Trojans," *Proc. 22nd Int. Conf. VLSI Design*, New Delhi, India, pp. 327–332, Jan. 2009.

R. Karri, J. Rajendran, K. Rosenfeld, and M. Tehranipoor, "Trustworthy hardware: identifying and classifying hardware trojans," *Computer*, vol. 43, no. 10, pp. 39–46, 2010.

S. Bhunia *et al.*, "Protection against hardware trojan attacks: towards a comprehensive solution," *IEEE Design & Test*, vol. 30, no. 3, pp. 6–17, 2013.

S. P. Mohanty, *Nanoelectronic Mixed-Signal System Design*. New York, NY, USA: McGraw-Hill Educ., 2015.

J. Rajendran, H. Zhang, O. Sinanoglu, and R. Karri, "High-level synthesis for security and trust," *Proc. IEEE 19th Int. On-Line Test. Symp. (IOLTS)*, Chania, Greece, pp. 232–233, July 2013.

X. Cui, K. Ma, L. Shi, and K. Wu, "High-level synthesis for run-time hardware Trojan detection and recovery," *Proc. 51st Annu. Design Autom. Conf.*, San Francisco, CA, USA, pp. 1–6, 2014.

D. McIntyre, F. Wolff, C. Papachristou, S. Bhunia, and D. Weyer, "Dynamic evaluation of hardware trust," *Proc. IEEE Int. Workshop HOST*, San Francisco, CA, USA, pp. 108–111, 2009.

S. P. Mohanty, N. Ranganathan, and V. Krishna, "Datapath scheduling using dynamic frequency clocking," *Proc. IEEE Comput. Soc. Annu. Symp. VLSI*, Pittsburgh, PA, USA, pp. 58–63, 2002.

S. P. Mohanty, M. Gomathisankaran, and E. Kougianos, "Variability aware architecture level optimization techniques for robust nanoscale chip design," *Comput. Elect. Eng*, vol. 40, no. 1, pp. 168–193, 2014.

X. Wang, M. Tehranipoor, and J. Plusquellic, "Detecting malicious inclusions in secure hardware: challenges and solutions," *Proc. IEEE Int. Workshop Hardw. Oriented Security Trust (HOST)*, Anaheim, CA, USA, pp. 15–19, June 2008.

M. Tehranipoor, and F. Koushanfar, "A survey of hardware trojan taxonomy and detection," *IEEE Design & Test of Computers*, vol. 27, no. 1, pp. 10–25, 2010.

A. Sengupta, and S. Bhadauria, "Automated exploration of datapath in high level synthesis using temperature dependent bacterial foraging optimization algorithm," *Proc. IEEE 27th Canadian Conf. Elect. Comput. Eng. (CCECE)*, Toronto, ON, Canada, pp. 1–5, May 2014.

C. Haubelt, T. Schlichter, J. Keinert, and M. Meredith, "Systemcodesigner: automatic design space exploration and rapid prototyping from behavioral models," *Proc. 45th ACM/IEEE Design Autom. Conf. (DAC)*, Anaheim, CA, USA, pp. 580–585, June 2008.

A. Sengupta, and R. Sedaghat, "Integrated scheduling, allocation and binding in high level synthesis using multi structure genetic algorithm based design space exploration," Proc. 12th Int. Symp. Qual. Electron. Design (ISQED), Santa Clara, CA, USA, pp. 1–9, Mar. 2011.

V. Krishnan, and S. Katkoori, "A genetic algorithm for the design space exploration of datapaths during high-level synthesis," *IEEE Transactions on Evolutionary Computation*, vol. 10, no. 3, pp. 213–229, 2006.

F. Ferrandi, P. L. Lanzi, D. Loiacono, C. Pilato, and D. Sciuto, "A multi-objective genetic algorithm for design space exploration in high-level synthesis," *Proc. IEEE Comput. Soc. Annu. Symp. VLSI (ISVLSI)*, Montpellier, France, pp. 417–422, 2008.

Z. Zeng, R. Sedaghat, and A. Sengupta, "A framework for fast design space exploration using fuzzy search for VLSI computing architectures," *Proc. IEEE Int. Symp. Circuits Syst. (ISCAS)*, Paris, France, pp. 3176–3179, May 2010.

C. Mandal, P. Chakrabarti, and S. Ghose, "GABIND: a GA approach to allocation and binding for the high-level synthesis of data paths," *IEEE Transactions on Very Large Scale Integration (VLSI) Systems*, vol. 8, no. 6, pp. 747–750, 2000.

D. S. H. Ram, M. C. Bhuvaneswari, and S. S. Prabhu, "A novel framework for applying multiobjective GA and PSO based approaches for simultaneous area, delay, and power optimization in high level synthesis of datapaths," *VLSI Design*, Art. no. 2, Jan. 2012.

Y. Liu, and K. M. Passino, "Biomimicry of social foraging bacteria for distributed optimization: models, principles, and emergent behaviors," *Journal of Optimization Theory and Applications*, vol. 115, no. 3, pp. 603–628, 2002.

K. Deb, and D. Kalyanmoy, *Multi-Objective Optimization Using Evolutionary Algorithms*. New York, NY, USA: Wiley, 2001.

S. Das, A. Biswas, S. Dasgupta, and A. Abraham, "Bacterial foraging optimization algorithm: theoretical foundations, analysis, and applications," in *Foundations of Computational Intelligence Volume 3* (Studies in Computational Intelligence), vol. 203, A. Abraham, A.-E. Hassanien, P. Siarry, and A. Engelbrecht, (eds.) Heidelberg, Germany: Springer, 2009, pp. 23–55.

A. J. Hu, "Formal hardware verification with BDDs: an introduction," *Proc. IEEE Pac. Rim Conf. Commun. Comput. Signal Process. 10 Years PACRIM 1987-1997 Netw. Pac. Rim*, vol. 2., Victoria, BC, Canada, pp. 677–682, Aug. 1997.

M. Bushnell, and V. Agrawal, *Essentials of Electronic Testing for Digital, Memory and Mixed-Signal VLSI Circuits*. New York: Springer, 2013.

H. Salmani, M. Tehranipoor, and J. Plusquellic, "New design strategy for improving hardware Trojan detection and reducing trojan activation time," *Proc. IEEE Int. Workshop Hardw. Oriented Security Trust (HOST)*, San Francisco, CA, USA, pp. 66–73, July 2009.

D. Agrawal, S. Baktir, D. Karakoyunlu, P. Rohatgi, and B. Sunar, "Trojan detection using IC fingerprinting," *Proc. IEEE Symp. Security Privacy (SP)*, Berkeley, CA, USA, pp. 296–310, May 2007.

S. Narasimhan *et al.*, "Multiple-parameter side-channel analysis: a noninvasive hardware trojan detection approach," in *Proc. IEEE Int. Symp. Hardw. Oriented Security Trust (HOST)*, Anaheim, CA, USA, June 2010, pp. 13–18.

Y. Jin, and Y. Makris, "Hardware Trojan detection using path delay fingerprint," Proc. IEEE Int. Workshop Hardw. Oriented Security Trust (HOST), Anaheim, CA, USA, pp. 51–57, June 2008.

X. Wang, H. Salmani, M. Tehranipoor, and J. Plusquellic, "Hardware Trojan detection and isolation using current integration and localized current analysis," *Proc. IEEE Int. Symp. Defect Fault Tolerance VLSI Syst. (DFTVS)*, Boston, MA, USA, pp. 87–95, Oct. 2008.

A. Sengupta, and S. Bhadauria, "Untrusted third party digital IP cores: power-delay trade-off driven exploration of hardware Trojan secured datapath during high level synthesis," *Proc. 25th Edition Great Lakes Symp. VLSI (GLSVLSI)*, Pittsburgh, PA, USA, pp. 167–172, 2015.

A. Sengupta, and R. Sedaghat, "Swarm intelligence driven design space exploration of optimal k-cycle transient fault secured datapath during high level synthesis based on user power-delay budget," Microelectronics Reliability, vol. 55, no. 6, pp. 990–1004, 2015. [Online]. Available: http://www.sciencedirect.com/science/article/pii/S0026271415000608

A. K. Kumar, D. Somasundareswari, V. Duraisamy, and M. Pradeepkumar, "Low power multiplier design using complementary pass-transistor asynchronous adiabatic logic," *International Journal of Computer Science and Engineering*, vol. 2, no. 7, pp. 2291–2297, 2010.

J. Crop, S. Fairbanks, R. Pawlowski, and P. Chiang, "150mV subthreshold asynchronous multiplier for low-power sensor applications," in *Proc. Int. Symp. VLSI Design Autom. Test (VLSI-DAT)*, Hsinchu, Taiwan, pp. 254–257, 2010.

Chapter 8

Trojan-resistant IP designs and isolation strategies

Anirban Sengupta[1]

This chapter exposes the readers to a new dimension of the hardware Trojan class known as the functional Trojan, which may be present during high-level synthesis (HLS) design. More specifically, the chapter deals with an elaborate discussion on how Trojan-resistant intellectual property (IP) designs can be generated, which includes functional HLS Trojan isolation techniques using redundancy logic. The chapter also presents an analysis of case studies along with their comparison with similar techniques.

8.1 Introduction

The evolution of modern computing systems has been marked by an increasing demand for application-specific computing systems that deliver superior performance, energy efficiency, and scalability. With the exponential growth of data-intensive and computation-heavy applications, ranging from artificial intelligence and machine learning to scientific modeling and multimedia processing, the need for customized hardware solutions has become indispensable. Such tailored designs enable developers to achieve optimized resource utilization, lower latency, and enhanced throughput compared to general-purpose processors (Pilato *et al.*, 2018).

The high-level synthesis (HLS) framework has emerged as a key enabler in this paradigm shift toward specialized hardware design. By allowing designers to describe functionality at a higher level of abstraction, HLS significantly shortens the design cycle and reduces the likelihood of human error inherent in traditional register-transfer level (RTL) design methodology. Moreover, the inclusion of the design space exploration (DSE) technique within the HLS framework facilitates systematic trade-offs among performance metrics such as area, power consumption, latency, etc. Through multi-objective optimization, DSE assists in identifying design configurations that best align with application-specific/designer/user constraints and performance goals (Anshul and Sengupta, 2023a; Schafer and Wang, 2020).

[1]Department of Computer Science and Engineering, Indian Institute of Technology Indore, India

The output of the HLS process, often realized as an intellectual property (IP) core, forms the backbone of custom computing systems used in diverse domains. These IPs implement crucial algorithms like discrete cosine transform (DCT), discrete wavelet transform (DWT), or finite impulse response (FIR) filtering, which are integral to real-time image, signal, and video processing (Mohanty and Meher, 2016). The increasing reliance on third-party IP (3PIP) vendors and offshore fabrication units, however, introduces a new dimension of security challenges in the global hardware design supply chain (Rathor *et al.*, 2024). While the use of third-party IPs enhances design modularity and accelerates time-to-market, it also exposes hardware systems to malicious threats such as IP piracy, counterfeiting, and the insertion of hardware Trojans. These Trojans, malicious modifications embedded within the design, can compromise functionality, leak sensitive information, or even disable the system under specific conditions. Such threats have raised significant concerns regarding the trustworthiness and integrity of outsourced hardware components (Rostami *et al.*, 2018; Sengupta *et al.*, 2024).

Existing security measures, including hardware watermarking (Anshul and Sengupta, 2023b) and biometric-based security techniques (Sengupta and Rathor, 2021; Sengupta and Rathor, 2020; Chaurasia *et al.*, 2022), primarily focus on detecting counterfeit IPs rather than ensuring Trojan-resistant design integrity. Furthermore, traditional post-fabrication detection mechanisms, such as side-channel analysis, formal verification, and simulation-based functional testing (Sengupta, 2017a, 2017b; Fyrbiak *et al.*, 2019), often fail to guarantee complete Trojan isolation, especially when an adversary implants stealthy modifications at the HLS tool library or foundry level.

Consequently, there was a growing need to shift from purely detective security mechanisms toward preventive and resilient design strategies. Developing Trojan-resistant IP architecture and a robust isolation framework during the early stages of the design process is critical to ensuring hardware trust. Such strategies should integrate seamlessly within existing HLS and DSE environments, providing designers with the ability to generate optimized, secure, and trustworthy IP cores without sacrificing performance or efficiency.

This chapter delves into the principles, challenges, and methodologies underlying Trojan-resistant IP design and isolation strategies (Sengupta *et al.*, 2023). It explores how modern hardware design frameworks can be extended to embed resilience against functional hardware Trojans while maintaining optimal performance through intelligent design space exploration. The discussion emphasizes Trojan-resistant architectural design, a redundancy-based approach, and a secure design flow that collectively ensure reliable and trustworthy system-on-chip (SoC) implementations in the presence of an untrusted design entity.

8.2 What is the need for designing optimal Trojan-resistant hardware IP?

Modern application-specific computing platforms rely extensively on reusable hardware IP cores to implement complex algorithms for digital signal, image, and video processing. These IP cores form the computational backbone of systems performing tasks such as image compression, audio filtering, and wireless communication. Examples include the discrete cosine transform (DCT) and Haar wavelet transform (HWT) used in image and video compression, finite impulse response (FIR) and infinite impulse response (IIR) filters applied in audio equalization and signal enhancement, and fast Fourier transform (FFT) modules employed in digital video broadcasting. Similarly, the JPEG-CODEC, a widely used image compression engine, is an essential IP component in digital cameras and multimedia devices. Owing to their widespread adoption, the reliability and integrity of such IP cores are paramount for ensuring the dependability of custom computing systems (Sengupta *et al.*, 2023; Schneiderman, 2010).

However, as hardware design and manufacturing increasingly depend on third-party IP vendors and offshore foundries, the risk of malicious hardware modifications, commonly known as *hardware Trojans*, has grown substantially. These Trojans are stealthy and often activated under specific trigger conditions, making them extremely difficult to detect using traditional functional or side-channel testing. Once inserted, they can compromise the security, performance, or even the safety of the end system. In safety-critical or mission-critical domains, such as defense electronics, healthcare instrumentation, and autonomous vehicles, a single compromised IP core can lead to catastrophic outcomes, data breaches, or irreversible system failures. Therefore, building inherent *Trojan resistance* for hardware IPs is a crucial design objective (Sengupta *et al.*, 2025).

Designing optimal Trojan-resistant hardware IPs serves several essential purposes. It ensures *functional reliability* by enabling the system to continue generating correct outputs even if a portion of the hardware becomes compromised. By integrating detection and isolation mechanisms within the IP design, such designs can identify anomalies and restrict the influence of infected components. Even if a single module is infected, the overall system functionality remains unaffected due to built-in corrective mechanisms that maintain operational integrity. Another compelling motivation lies in *consumer safety and trust*. Devices powered by application-specific SoCs often operate in close interaction with human environments, whether through communication systems, biomedical devices, or automotive control units. Trojan-induced malfunctions in such systems can have direct consequences on human safety, making trust assurance in hardware design indispensable. Moreover, in large-scale commercial deployments, undetected Trojans can cause severe financial and reputational losses, reinforcing the importance of preventive architectural strategies (Sengupta *et al.*, 2023).

In summary, the need for optimal Trojan-resistant hardware IP design arises from the intersection of three critical challenges: ensuring the reliability of high-performance computing systems, safeguarding the hardware supply chain against malicious effects, and maintaining consumer trust in the integrity of modern electronics. As the complexity of IP-based SoCs continues to rise, embedding security and robustness into the design process itself is the most effective strategy to design trustworthy and dependable hardware systems for future computing applications.

8.3 Salient features of the chapter

This chapter highlights key methodologies in the optimized design of a Trojan-resistant hardware IP core using the HLS framework (Sengupta *et al.*, 2023). The salient features of the chapter are summarized as follows:

- Integration of security and optimization in the HLS framework
  - This chapter emphasizes embedding Trojan detection and isolation capabilities directly within the design process, along with the optimization framework, thereby transforming the traditional HLS framework into a *low-cost security-aware synthesis framework.*
- Exploration of optimized Trojan-resistant hardware architectures using particle swarm optimization (PSO)-based design space exploration (DSE)
  - This chapter demonstrates how a particle swarm optimization (swarm intelligence) algorithm can effectively guide the DSE process to identify an optimal Trojan-resistant hardware IP core configuration.
  - The exploration process balances performance objectives such as area and delay (latency) while ensuring that the selected architecture maintains robust Trojan resistance capability.
- Trojan-resistant design flow for reusable IP cores
  - A *Triple Modular Redundancy (TMR)*-based design strategy (Sengupta *et al.*, 2023) is discussed for achieving both detection and isolation of hardware Trojans within reusable IP cores.
  - The TMR approach enhances reliability by allowing the system to continue functioning correctly even if one or more modules become compromised.
- Distinct multi-vendor allocation policy for enhanced security
  - The chapter discusses a *multivendor IP allocation policy* (Sengupta *et al.*, 2023) that strategically assigns redundant units of the TMR to different, independent IP vendors.
  - This policy significantly reduces the probability of collusion between distinct vendors for the same Trojan payload insertion across identical design units, strengthening the overall trustworthiness of the final generated IP.
- Comprehensive framework for reliable IP/system-on-chip (SoC) design
  - The discussed design methodology is applicable to a wide range of digital signal processing (DSP) and computation-intensive IP cores, such as DCT, FIR, FFT, JPEG-CODEC, etc.
  - By combining performance optimization with embedded security mechanisms, the framework supports the development of *scalable, reliable, and trustworthy SoC solutions* for multimedia and computing applications.

8.4 State-of-the-art

Research in the domain of hardware security has extensively focused on securing integrated circuits and IP cores from malicious modification. Traditional

HLS-based IP/SoC design methodologies have primarily emphasized area-delay optimization or security-aware synthesis exclusively. However, integrating Trojan resistance within the optimization process of HLS, especially while balancing multiple design constraints such as latency and area, poses substantial challenges. Therefore, this chapter discusses the design process of *optimal Trojan-resistant hardware IPs* using the HLS framework (Sengupta *et al.*, 2023). Generally, Trojan detection mechanisms typically target behavioral anomalies or side-channel signatures. However, in the context of HLS functional Trojans, resistance and isolation require incorporating multi-vendor scheduling and redundancy strategies within the HLS design flow (Sengupta *et al.*, 2023). Such integration is inherently sophisticated, as it demands the concurrent handling of DSE and a secure scheduling strategy.

Some methodologies (Sengupta *et al.*, 2017; Sengupta and Mohanty, 2019) have made progress toward security-aware scheduling at the HLS level, but these methods largely focused on Trojan detection without guaranteeing full resistance or isolation capability. These works were limited to low-cost, security-aware DSP IPs that could only flag suspicious activity rather than sustain reliable operation under potential Trojan infection. Further, in the broader context of optimization, several DSE approaches have been developed to balance multiple performance metrics. Genetic algorithm-based DSEs have been discussed to identify Pareto-optimal system-level configurations (Liao *et al.*, 2023; Qian *et al.*, 2023), while machine learning-assisted (Liu *et al.*, 2013) and simulated annealing-based (Williams *et al.*, 2008) approaches have been applied for exploring power-performance trade-offs. Other works, such as the bacterial foraging-based (Sengupta and Bhadauria, 2015) framework, have attempted to reduce design cost in fault-tolerant systems. These optimization strategies lack integration with the HLS Trojan-resistance mechanism and are incapable of isolating functional hardware Trojans in IP cores.

Beyond design-level optimization, several studies have discussed Trojan detection-oriented frameworks using property checking, code coverage, or equivalence verification. For instance, the code-coverage-based method (Hu, 1997) identifies inactive or stable signals during testing as potential Trojan carriers, while equivalence analysis (Bushnell and Agrawal, 2001) reduces the set of suspicious nodes. Although such approaches enhance test efficiency, they incur high runtime overheads and remain purely *detective* in nature, offering no inherent isolation. Similarly, system-level techniques for Trojan detection (Tehranipoor and Koushanfar, 2010) and circuit-level strategies employing concurrent error detection or supply current monitoring (Agrawal *et al.*, 2007; Narasimhan *et al.*, 2010; Wang *et al.*, 2008; Rajendran *et al.*, 2016) improve observability but fail to maintain functionality in the presence of active Trojans. Other state-of-the-art works explored approximate circuit-based defenses (Martin *et al.*, 2018), obfuscated TMR (Siboni and Shekarian, 2019), and Trojan neutralization in supervisory control systems (Gunti and Lingasubramanian, 2017). While effective in their respective contexts, these approaches were designed for gate-level or control-level protection and do not address functional HLS Trojan isolation in DSP-based hardware IP cores. Similarly, hardware camouflage strategies (Li *et al.*, 2021) have been used to hide critical logic components but lack integration with multi-vendor design policy or a high-level optimization framework.

Recent advances in hardware verification and machine learning have also been applied to Trojan detection. For example, equivalence checking techniques using finite-state machine datapaths (Abderehman *et al.*, 2022a), machine learning-based detection (Kulkarni *et al.*, 2016), discrepancy analysis (Fezzardi *et al.*, 2018), and a C/C++ to RTL equivalence checking framework (Abderehman *et al.*, 2022b) provide incremental improvements in Trojan detection. Moreover, assertion-based verification frameworks (Fern and Cheng, 2019; Alsaiari and Gebali, 2019) further automate the identification of design blind spots. Nevertheless, all these methods remain limited to post-design or runtime detection, lacking the architectural resilience required to isolate or neutralize functional HLS Trojans.

In summary, all approaches discussed above predominantly address Trojan detection. However, Sengupta *et al.* (2023) discussed in this chapter distinguish themselves by integrating PSO-driven design space exploration with a TMR-based multi-vendor allocation policy, resulting in optimized, low-cost, and fully *HLS Trojan-resistant hardware IP designs* capable of isolating malicious units while ensuring correct IP design functionality.

8.5 Threat model

In the modern semiconductor ecosystem, hardware designs often traverse multiple untrusted environments during their development, testing, and integration phases. This chapter focuses on the threat model associated with *functional hardware Trojans* that are stealthily embedded in 3PIP cores. These IP cores, widely utilized to accelerate design cycles and reduce development costs, are often sourced from multiple vendors whose trustworthiness may vary. When a malicious actor modifies an IP core during the design process, the resulting Trojan can alter functional behavior (Bhunia *et al.*, 2014; Hu *et al.*, 2021). The stealthy nature of such Trojans, remaining dormant until a rare trigger condition occurs, renders conventional design verification and testing methods largely ineffective. From the *attacker's perspective*, the motivations behind introducing functional Trojans can vary significantly. A malicious vendor might attempt to:

(i) Damage the reputation of a competing firm by introducing reliability issues that appear to stem from design flaws, thereby affecting market confidence.
(ii) Sabotage mission or safety-critical systems by inducing malfunctions that can compromise the entire infrastructure.

In this model, the IP designer, responsible for integrating the overall design, is assumed to be trusted. However, the third-party HLS tool developer/vendor house provides a built-in HLS tool library that contains micro-3PIP modules/components, which are potentially untrustworthy. The lack of visibility into third-party HLS tools and their respective micro-3PIP module library creates a major blind spot for integrators, complicating both Trojan detection and isolation efforts. Furthermore, since the detailed internal structure and source code of micro-3PIP modules are typically inaccessible to integrators, detection using post-design validation techniques becomes challenging. Functional Trojans can exploit this opacity to evade detection throughout standard verification flows. Once integrated into the SoC hardware system, these Trojans can jeopardize system-level reliability, leading to potentially catastrophic consequences in application-specific integrated circuits (ASICs) and real-time embedded systems.

This threat model underscores the need for robust hardware IP design that can ensure Trojan resistance without requiring full transparency from micro-3PIP vendors. By assuming that untrusted micro-3PIPs may carry hidden malicious logic, IP designers must employ methods that can isolate, analyze, and validate each IP core's functional trustworthiness before deployment.

8.6 Real-life case studies

The practical impact of functional hardware Trojans (FHT) extends far beyond theoretical analysis. Their infiltration into real-world digital signal processing (DSP), machine learning (ML), and embedded systems can cause severe functional deviations, reliability issues, and even safety-critical failures. The following case studies illustrate how the lack of Trojan resistance in IP cores can lead to adverse real-life consequences across diverse computing domains.

a) Convolutional Neural Network (CNN) accelerators

   CNNs are fundamental to modern image classification, object detection, and biometric recognition systems. The convolution layer IP (Sengupta *et al.*, 2021) used in CNN accelerators computes output feature maps by sliding a filter kernel over the input image. If a backdoor functional Trojan is embedded in the arithmetic or logic units of this convolution core, it can alter output pixel computations, leading to misclassifications or detection errors. For instance, in a facial recognition system, even a minor alteration in the convolution output can cause authentication failure or unauthorized access. In medical imaging, such as MRI or CT scan analysis, a compromised convolution filter could distort diagnostic features, leading to incorrect medical interpretations and jeopardizing patient safety. This demonstrates how undetected functional Trojans can undermine both security and ethical reliability in AI-driven healthcare systems.

b) Image compression and multimedia processing

   In image and video compression systems, such as those employing JPEG or MPEG standards, the discrete cosine transform (DCT) core plays a vital role in computing transformed pixel values. A malicious modification in this core can lead to the generation of corrupted compressed data. While such corruption might be visually subtle, its effects can be devastating in teleradiology and telemedicine, where diagnostic imaging relies on compressed medical data transmission. If a functional Trojan manipulates compression ratios or quantization parameters, critical diagnostic details could be lost, potentially resulting in fatal misdiagnoses. Therefore, functional integrity within DCT-based DSP IPs is essential for medical and industrial imaging systems.

c) Machine learning coprocessors

Hardware Trojans also pose a significant risk to machine learning accelerators and coprocessor systems. An attacker with access to an ML inference core could implant a backdoor Trojan that subtly alters prediction outcomes once a specific trigger condition is met. For example, in automated healthcare devices, such tampering could misclassify medical symptoms, influencing drug prescriptions or diagnosis outcomes to benefit malicious stakeholders, such as those seeking profit through biased or erroneous recommendations. Similarly, compromised ML coprocessors in autonomous vehicles could misinterpret sensor data, leading to dangerous navigation decisions, etc. Apart from them, as ML systems have become the core components of many modern digital electronics systems, the presence of hardware functional Trojans can lead to multiple negative consequences.

d) Digital filter and edge detection cores

Beyond CNNs and DCTs, several other digital filter IP cores, including infinite impulse response (IIR) filters, Gaussian filters, blur filters, sharpening filters, Laplacian and Sobel edge detectors, are widely used in consumer electronics, image enhancement, and modern industrial systems. A functional Trojan embedded in these cores could degrade image quality, produce false edges, or distort visual features. In surveillance or defense systems, such distortions could suppress critical details, leading to false security alerts or failure to detect intrusions. Thus, maintaining the functional trust of such cores is paramount for ensuring accurate visual computation in both civilian and defense-grade applications.

Therefore, these real-world scenarios emphasize the necessity of Trojan-resistant IP design methodologies. When preventive measures fail, the system architecture must at least possess the ability to contain or isolate compromised functional units to maintain correct overall behavior. Integrated techniques such as *triple modular redundancy (TMR)* and *distinct multi-vendor allocation policy* can be employed to detect inconsistencies between replicated modules and quarantine the Trojan-affected components (Sengupta *et al.*, 2023). By combining design-level resistance with isolation-based recovery strategies, hardware systems can sustain reliable functionality even in the presence of sophisticated functional Trojan attacks.

8.7 Details of the Trojan resistance methodology

8.7.1 Key considerations of the methodology (Sengupta et al.*, 2023)*

The Trojan-resistant methodology (Sengupta *et al.*, 2023) for developing Trojan-resistant IP designs is built on a set of critical design considerations that ensure robustness, functional reliability, and resistance to malicious intrusions. These considerations define the operational assumptions and design environment under which the methodology (Sengupta *et al.*, 2023) achieves effective Trojan detection and isolation.

a) Fault-tolerant voter design

   At the core of the TMR-based Trojan-resistant IP design approach lies a *fault-tolerant voter*, which serves as the final decision-making unit responsible for generating the correct functional output from multiple replicated modules. The voter architecture, inspired by existing resilient design frameworks (Kshirsagar and Patrikar, 2009), is engineered to remain operationally accurate even in the presence of transient faults or localized hardware disturbances. By ensuring that the voter itself is immune to failure, the methodology guarantees that the system's majority decision remains trustworthy, thereby upholding the integrity of the overall IP design.

b) Error detection and reliability mechanisms

   To further enhance design robustness, a dedicated error detection block (EDB) is incorporated. This multi-stage module continuously monitors comparator outputs and functional discrepancies among redundant units. Based on Kshirsagar and Patrikar (2009) and Kachave and Sengupta (2016), the EDB identifies abnormal functional deviations that may result from hidden Trojan activity or comparator fault. Through layered fault detection and validation stages, it provides an additional safeguard against the silent propagation of errors, ensuring that compromised outputs are flagged or isolated before they affect IP design functionality.

c) Trusted components and design hierarchy

   A key methodological assumption is that certain system components, specifically, the *voter* and *error detection block*, are *trusted* and *Trojan-free*. These modules are assumed to be developed *in-house* by the IP designer, a trusted entity within the design ecosystem. This hierarchical trust boundary limits exposure to external manipulation and ensures that critical control logic remains secure from third-party tampering.

d) Vendor isolation and information confidentiality

   The approach (Sengupta *et al.*, 2023) also emphasizes isolation of functional HLS Trojans. Each micro-3PIP core is sourced from an independent IP vendor, and information regarding the identity, function, and design of other IP vendors remains confidential and known only to the IP designer (who is responsible for integration using TMR logic). The micro-3PIP vendors are not cognizant of each other; therefore, there is no chance of collusion and a common Trojan payload. This strategic information separation minimizes the probability of vendor collusion, thereby reducing the likelihood of coordinated Trojan insertions with consistent payloads across multiple IPs (Sengupta *et al.*, 2017).

e) Assurance of minimum Trojan detection capability

   Given these considerations, the Trojan-resistant methodology (Sengupta *et al.*, 2023) discussed in this chapter minimally guarantees a *Trojan detection capability* even in complex or partially compromised environments.

The combined use of *fault-tolerant voting*, *multi-stage error detection*, and *vendor isolation* ensures that any Trojan-induced anomaly within the redundant logic will be detected, contained, or outvoted by trusted modules. This design philosophy enables the hardware IP design to sustain correct functional behavior even in the presence of untrusted or potentially compromised micro-3PIP modules.

8.7.2 Definition of the problem addressed

The primary problem addressed in this chapter revolves around designing hardware IP cores that are resistant to functional hardware Trojans while maintaining optimal design efficiency and performance. Formally, the problem can be expressed as the task of developing a Trojan-resistant IP core design that minimizes the *hybrid design cost*, defined as a combination of two key performance metrics:

(i) Total TMR area overhead (A_{TMR}) incurred due to redundancy mechanisms such as triple modular redundancy, and
(ii) Total execution time (T_{TMR}) associated with the TMR architecture.

The objective is to achieve an *optimal trade-off* between hardware security and resource efficiency through DSE that determines the best resource configuration set $\{P_i\}$ for a given IP design. This problem specifically focuses on functional Trojans. The scope of the addressed problem does not extend to other classes of hardware Trojans. Furthermore, the vulnerability extends to the *micro-3PIPs* embedded within commercial *HLS tool libraries*, which are often sourced from multiple vendors. Any untrustworthy or compromised library element can introduce a Trojan-infected module into the synthesized IP core. When such compromised IPs are integrated into larger custom computing systems, they can result in malfunctioning devices, degraded performance, or safety-critical operational hazards for the end user.

8.7.3 Models used for assessment of the IP design

To evaluate the effectiveness of the Trojan-resilient hardware IP architecture, the analytical models used (Sengupta *et al.*, 2023) are described below. The assessment framework primarily includes three metrics: area estimation, latency calculation, and the fitness value/design cost function.

a) Area evaluation model

The *area metric* represents the total resource consumption of the TMR design. It considers both the type and the quantity of hardware components employed from each vendor. The total occupied area A_{TMR} is calculated as follows:

$$A_{TMR} = \sum_{i=1}^{2} \sum_{j=1}^{3} \left(A_{r_i}^{(v_j)} \times N_{r_i}^{(v_j)} \right) \tag{8.1}$$

Where, $A_{r_i}^{(v_j)}$= area utilized by a single resource type r_i from vendor v_j, and $N_{r_i}^{(v_j)}$= number of instances of that resource type used from the same vendor.

b) Latency/Delay evaluation model

The *delay model* measures the overall execution latency of the IP design. It accounts for both the number of control steps during scheduling and the delay of each functional unit in that particular control step. The total design delay D_{TMR} is expressed as:

$$D_{TMR} = (k_m \times t_m) + (k_a \times t_a) \tag{8.2}$$

Where, k_m = number of control steps required by multiplier, k_a = number of control steps required by the adder, t_m = propagation delay of a single multiplier, and t_a = propagation delay of a single adder.

c) Fitness/Design cost function

The *design cost function* serves as a fitness model of the PSO-DSE process. It combines normalized versions of the total area and total delay, weighted by user-defined weighing factors. The cost function C_{design} is defined as:

$$C_{design} = \alpha_1 \left(\frac{A_{TMR} - A_{Con}}{A_{\max}} \right) + \alpha_2 \left(\frac{D_{TMR} - D_{Con}}{D_{\max}} \right) \tag{8.3}$$

Where, α_1, α_2 = weighting factors representing the designer's emphasis on area or delay optimization, A_{TMR}, D_{TMR} = computed area and delay of the proposed Trojan-resistant IP design, and A_{max}, D_{max} = maximum area and delay. Together, these three models form a comprehensive evaluation framework to analyze the area, delay, and cost of the low-cost Trojan-resistant IP design.

8.7.4 Optimal Trojan-resistant TMR design

Figure 8.1 highlights the overview of the low-cost Trojan-resistant IP design framework (Sengupta *et al.*, 2023). As shown in Figure 8.1, the design methodology begins with the control data flow graph (CDFG) representation of the target hardware application. The framework takes as inputs the application's CDFG, initial PSO parameters (inertia weight, acceleration coefficients, population size, and termination conditions), a multi-vendor IP library, and designer-specified constraints. Each resource, such as adders and multipliers, is sourced from one of three independent vendors supplying micro-3PIPs.

As shown in Figure 8.1, the swarm intelligence (PSO)-based design space exploration algorithm explores various resource configurations to identify an optimal hardware IP architecture mapping that minimizes the hybrid cost function, comprising total area (A_{TMR}) and total execution time (T_{TMR}) of the Trojan-resistant IP design. Each particle in the PSO represents a unique configuration of resource allocation and scheduling. The iterative process updates local and global best solutions, ensuring an increasingly optimized search space. Moreover, mutation operations are introduced periodically to avoid premature convergence and improve the diversity of potential solutions. At the end of the exploration, the framework outputs an *optimal (low-cost) Trojan-resistant IP design* for the input application that exhibits the best balance between performance and security. Other

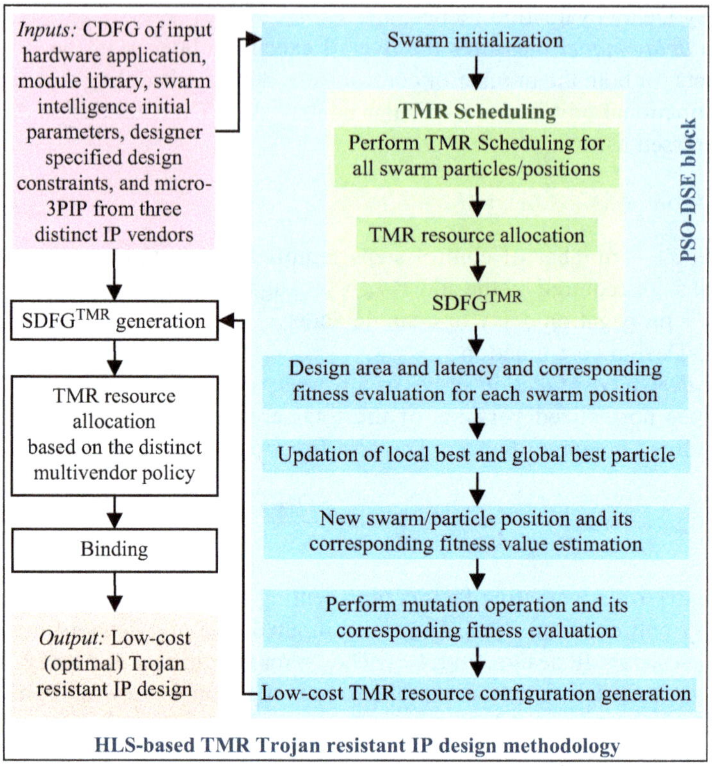

Figure 8.1 Overview of the HLS-based TMR Trojan-resistant methodology

major blocks (shown in Figure 8.1) include TMR scheduling and a multi-vendor allocation framework. The TMR scheduling framework performs scheduling of input CDFG based on PSO-initialized/generated resource configurations, and the multi-vendor allocation framework performs hardware (functional unit) allocation based on three distinct micro-3PIP vendors. The details of each step are discussed below.

Next, the *TMR architecture* forms the foundation of Trojan resistance. Figure 8.2 depicts the complete Trojan-resistant architecture (Sengupta *et al.*, 2023). As shown in Figure 8.2, each functional unit in the original design is replicated twice, producing three identical operational paths: original, duplicate, and triplicate. The results from these units are passed to a *fault-tolerant voter* (Kshirsagar and Patrikar, 2009), which determines the final IP design output based on the majority rule. This ensures that even if one of the units is compromised by a Trojan, the voter selects the correct result from the two unaffected modules. The 4:1 multiplexer (MUX) output facilitates Trojan detection when the select inputs (S1, S0) are set to 01, 10, or 11, while it offers Trojan resistance when the select inputs are 00. To guarantee that the design consistently achieves at least Trojan

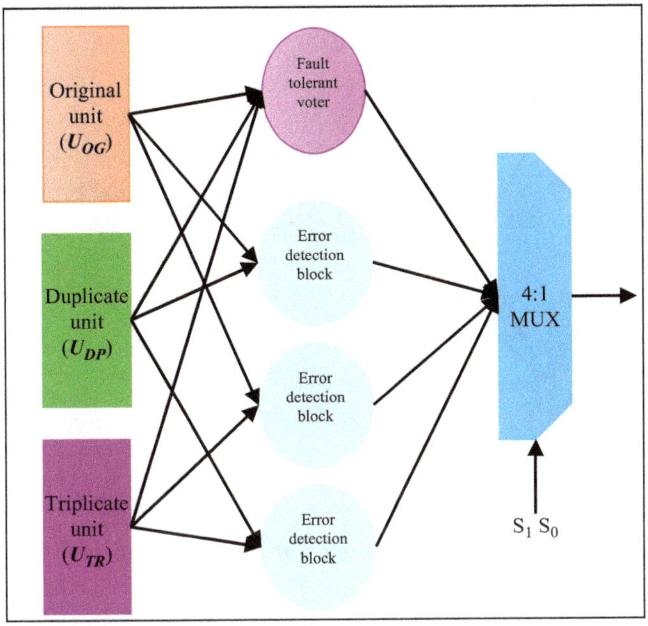

Figure 8.2 Demonstration of the TMR Trojan resistance framework

detection capability, a multi-stage configuration incorporating error detection blocks (EDBs) has also been integrated (Sengupta *et al.*, 2023). Figure 8.3 illustrates the EDB architecture used in the Trojan-resistant framework (Sengupta *et al.*, 2023). The EDB was adopted from Kshirsagar and Patrikar (2009) and Kachave and Sengupta (2016), which monitors discrepancies between redundant unit outputs using multiple comparators and a secondary voter. If one comparator malfunctions or produces inconsistent results, the EDB's voter derives the correct majority outcome from the remaining comparators, thereby maintaining system reliability. The area overhead introduced by this additional logic remains negligible while substantially improving Trojan isolation capability (Sengupta *et al.*, 2023).

Next, a critical security enhancement in the methodology (Sengupta *et al.*, 2023) arises from the *distinct multi-vendor allocation policy*. In this scheme, each of the three TMR units (original, duplicate, and triplicate) exclusively utilizes resources from a *unique, distinct vendor*. For example, if vendor V_1 supplies resources to the original unit, vendors V_2 and V_3 are allocated to the duplicate and triplicate units, respectively. This strict separation significantly reduces the probability that all three redundant paths contain identical Trojan payloads, as vendors operate independently and remain unaware of each other's participation in the design process. If one vendor's module is compromised, the Trojan can only affect one of the three TMR units. The fault-tolerant voter then masks the erroneous output from an infected unit by selecting the majority output produced by the remaining two Trojan-free units. This isolation mechanism ensures functional

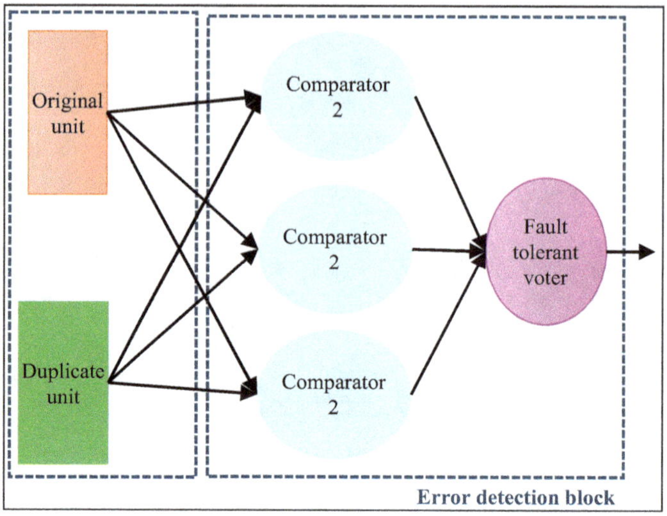

Figure 8.3 Error detection block adopted from Kshirsagar and Patrikar (2009)

correctness despite partial design contamination, thereby providing robust resistance against functional Trojans embedded at the vendor or HLS tool library level.

Following the resource allocation phase, the TMR CDFG (generated post replication of functional units) undergoes scheduling using a LIST scheduling algorithm. This scheduler prioritizes the execution of maximum operations per control step, considering both resource availability and data dependencies (Sengupta *et al.*, 2023). For each configuration explored during the PSO-DSE process, the total area and execution time are computed. The *fitness cost function*, representing a weighted hybrid of A_{TMR} and T_{TMR}, is evaluated iteratively until the termination criterion is met (Sengupta *et al.*, 2023). The optimal (low-cost) Trojan-resistant TMR architecture for the target IP core is designed using the global best solution obtained from this process. The details corresponding to the swarm intelligence-based DSE process are discussed in the next subsection.

8.8 Details of swarm intelligence and its role in designing low-cost Trojan-resistant IP design

8.8.1 Integration of PSO-DSE with Trojan-resistant design

The integration of swarm intelligence-based design space exploration with the Trojan-resistant design framework provides an efficient framework for developing secure and low-cost (optimal) hardware IP design. Figure 8.4 highlights the integration flow of PSO-DSE with TMR-trojan-resistant methodology. Swarm intelligence (PSO) is particularly suited for this task due to its inherent capability to balance global exploration and local exploitation during the optimization process. Unlike other metaheuristic-based DSE techniques such as genetic algorithms (GA)

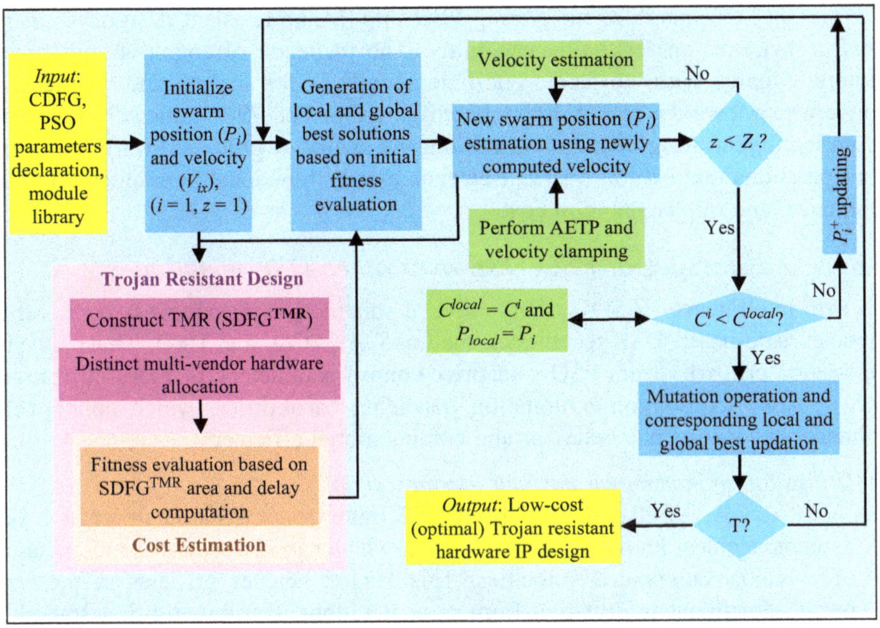

Figure 8.4 Flow chart of PSO-DSE process integrated with TMR methodology

(Krishnan and Katkoori, 2006), ant colony optimization (ACO) (Kao *et al.*, 2012), and bacterial foraging optimization (BFO) (Sengupta and Bhadauria, 2014), PSO exhibits faster convergence characteristics and reduced computational overhead, making it highly suitable for security-driven design environments where time and resource efficiency are critical (Mishra and Sengupta, 2014).

In the context of Trojan-resistant IP development, PSO aids in exploring a vast multidimensional design space characterized by security, latency, and area constraints. Each particle in the swarm represents a potential design configuration, while its movement through the search space is governed by both individual experience (cognitive component) and collective intelligence (social component). This cooperative learning mechanism enables PSO to efficiently locate optimal or near-optimal (low-cost) solutions that provide Trojan resistance without excessively compromising resource utilization. The ability of PSO to self-adjust exploration parameters based on design-space feedback ensures that the search does not prematurely converge on suboptimal solutions, a common issue observed in GA and ACO approaches (Mishra and Sengupta, 2014).

Furthermore, PSO offers a lightweight implementation framework with minimal control parameters and simpler mathematical modeling compared to other evolutionary algorithms. The design process benefits from dynamic adaptability and stability, particularly in scenarios with high variable dimensionality and nonlinear interdependencies. This adaptive control minimizes the likelihood of being trapped in local optima, a drawback frequently encountered in BFO due to its static step size and weakening bacterial interaction model.

Another advantage of integrating PSO into Trojan-resistant design lies in its flexible hyperparameter tuning capability. The inclusion of cognitive and social factors, coupled with stochastic coefficients, allows for smooth and progressive convergence toward a global optimal configuration. The PSO framework thus acts as an intelligent design navigator, guiding the synthesis process toward hardware configurations that exhibit a robust balance among functional correctness, Trojan resilience, and implementation cost.

8.8.2 Advantages of PSO-DSE over other DSE techniques

Swarm intelligence (PSO) offers several distinctive advantages over other metaheuristic-based DSE techniques such as GA, ACO, and BFO. These advantages stem primarily from PSO's adaptive control parameters, memory utilization, and dynamic exploration-exploitation balancing capabilities, which collectively enhance its convergence behavior and computational efficiency.

a) *Utilization of momentum through inertia weight*
   Unlike GA, ACO, and BFO-based DSE frameworks, PSO incorporates a key dynamic element known as *inertia weight*, which carries forward the momentum of previously computed velocities. This feature enables particles to preserve useful directional information from prior iterations, thereby ensuring smoother and more informed navigation through the search space. By contrast, algorithms that lack such momentum-based adaptation tend to exhibit erratic search behavior and are more susceptible to premature convergence within local optima. The inclusion of inertia weight thus enhances PSO's global search capability and accelerates convergence toward optimal Trojan-resistant design configuration.
b) *Dynamic balance between exploration and exploitation*
   PSO employs a linearly decreasing inertia weight strategy, typically ranging from 0.9 with initialization at 0.1. This dynamic adjustment allows the algorithm to take larger exploratory steps during early iterations, enabling a wide search across diverse design configurations, and gradually focus on fine-tuning solutions during later stages. Such controlled convergence behavior effectively balances exploration (diversification) and exploitation (intensification), ensuring the discovery of globally optimal solutions without unnecessary computational overhead. In contrast, conventional GA or BFO methods operate with fixed or less adaptive control parameters, leading to slower convergence and a higher likelihood of being trapped in suboptimal regions of the design space.
c) *Enhanced adaptability through hyperparameter integration*
   Another key advantage of PSO lies in its diverse set of hyperparameters, including random coefficients, inertia weight, and social and cognitive learning factors. These elements collectively guide the trajectory of each particle based on both its local best solution (P_{local}) and the global best solution (P_{global}) identified by the swarm. This dual feedback mechanism fosters robust convergence behavior, allowing the algorithm to dynamically adjust search trajectories in response to the evolving design landscape. Other DSE approaches, such as GA, ACO, and BFO, typically lack these features and thus require extensive parameter tuning to achieve comparable performance.

8.8.3 Encoding of PSO-DSE and its iterative procedure

The PSO-DSE process begins with the representation (encoding) of each particle as a potential architectural/resource configuration within the design space, where each dimension (Z) of the particle corresponds to a specific design resource, such as an adder and multiplier. In the encoding process, each particle is defined by two primary parameters: its *position* and *velocity*. The position vector represents the design configuration (i.e., the quantity or type of functional resources), whereas the velocity vector indicates the rate of change of these design parameters across successive iterations. The initial positions of the particles are strategically assigned to ensure a well-distributed exploration of the search space. The swarm particle positions are initialized as follows:

- The *first particle* is initialized with the minimum resource configuration:
- $P_1 = (\min(Z_1), \min(Z_2))$, representing a minimal design scenario, here Z corresponds to the dimension of that swarm position. *Note*: The minimum value corresponding to each dimension (i.e., functional resources) is available in the module library.
- The *second particle* is initialized with the maximum resource configuration:$P_2 = (\max(Z_1), \max(Z_2))$. *Note*: The maximum value corresponding to each dimension (i.e., functional resources) is available in the module library.
- The *third particle* is initialized as the average of the minimum and maximum configurations:

$$P_3 = \left(\frac{(\min(Z_1) + \max(Z_1))}{2}\right), \left(\frac{(\min(Z_2) + \max(Z_2))}{2}\right).$$

The remaining particles $(P_4, P_5, \ldots, P_n)$ are randomly initialized using:

$$P_{ik} = \frac{(w + s)}{2} \pm f \tag{8.4}$$

Here, w and s denote the minimum and maximum allowable functional resource limits, respectively, and f is a random number within the range $[w, s]$. Here, P_{ik} represents the configuration of the ith particle in the kth design dimension, which includes resource types such as adders or multipliers.

Iterative procedure: Once the swarm is initialized, each particle's fitness value is computed using a fitness/cost function (using (8.3)). The particle with the minimum cost function value is selected as the *global best* (P_{global}), while each particle maintains its own *local best* (P_{local}) configuration. In each iteration, the velocity and position of every particle are updated based on the following equations:

$$v_i^{(t+1)} = h \times v_i^{(t)} + c_1 \times r_1 \times \left(P_{\text{local}_i} - P_i^{(t)}\right) + c_2 \times r_2 \times \left(P_{\text{global}} - P_i^{(t)}\right) \tag{8.5}$$

$$P_i^{(t+1)} = P_i^{(t)} + v_i^{(t+1)} \tag{8.6}$$

Here, h is the inertia weight (linearly decreasing from 0.9 to 0.1), c_1 and c_2 denote the cognitive and social learning factors, respectively, and r_1 and r_2 are uniformly distributed random numbers in the range [0,1]. The velocity and position update mechanism allows each particle to adaptively refine its trajectory toward the global optimal solution. To prevent uncontrolled exploration, *velocity clamping* is employed [50], which restricts the maximum allowable velocity magnitude to ensure particles remain within the defined design space. If a particle's updated position exceeds the boundary limits, *adaptive end-terminal perturbation* (AETP) is applied to reposition it within the valid search boundaries (Mishra and Sengupta, 2014). This approach maintains search stability and ensures meaningful design exploration. Following each iteration, the cost function is recalculated for all particles. If a particle's current fitness value improves upon its previous, the local and global bests are updated accordingly. To enhance diversity and avoid premature convergence, a *mutation operation* is periodically introduced into the swarm (Mishra and Sengupta, 2014). These mutations introduce slight perturbations in the particles' positions, encouraging broader exploration and helping the swarm escape local minima.

Termination criteria: The iterative PSO process continues until a termination condition is met. The termination occurs either when:

- The algorithm converges to a global minimum and exhibits no further improvement in fitness values for ten consecutive iterations ($T_1 = 10$); or
- The maximum allowable iteration count is reached ($T_2 \leq 50$).

Upon termination, the particle corresponding to the global best configuration is declared as the *low-cost (optimal) Trojan-resistant IP architecture configuration*.

8.9 Example of Trojan-resistant IP design on DCT core

The low-cost TMR-based Trojan-resistant methodology implementation is demonstrated on a discrete cosine transform (DCT) application. The DCT core is extensively used in various signal and image processing applications, such as JPEG and MPEG compression standards. The initial step involves transforming the high-level algorithmic representation of the 4-point DCT into its corresponding control data flow graph (CDFG). The CDFG captures the computational dependencies among arithmetic operations by parsing the mathematical formulation of the DCT function. This CDFG serves as the foundational model for resource allocation, scheduling, and Trojan-resistant design exploration. Following CDFG generation, the *triple modular redundancy* technique is applied to generate the TMR CDFG. As discussed in previous sections, three identical computational units—original, duplicate, and triplicate—are generated. These units execute concurrently to ensure Trojan resistance. Next, the PSO-DSE algorithm determines the optimal resource configuration for TMR-resistant design, balancing latency, area, and security trade-offs. As discussed previously, the three DCT units are integrated using components sourced from three different micro-3PIP vendors, vendor V1, vendor V2, and vendor V3, respectively. This *distinct multi-vendor* resource allocation policy introduces isolation, thereby minimizing the likelihood of coordinated Trojan activation across the redundant units (collusion between the vendors).

Once the TMR-based CDFG is constructed, the design is scheduled using a *LIST scheduling* algorithm, ensuring efficient allocation of computational operations across available control steps. The final scheduled data flow graph (SDFG) represents the optimized execution sequence of arithmetic operations for the TMR DCT architecture. Finally, a *majority voting mechanism* is employed at the final stage to compare the outputs of the three units. In case of any mismatch, the output is considered potentially Trojan-infected, and the final output is decided based on the majority of the results. Figure 8.5. depicts the SDFG (TMR) of 4-point DCT with three multipliers (*) and one adder (+). The optimization process effectively reduces redundant hardware overhead compared to conventional static TMR implementation while maintaining functional Trojan immunity. Furthermore, the

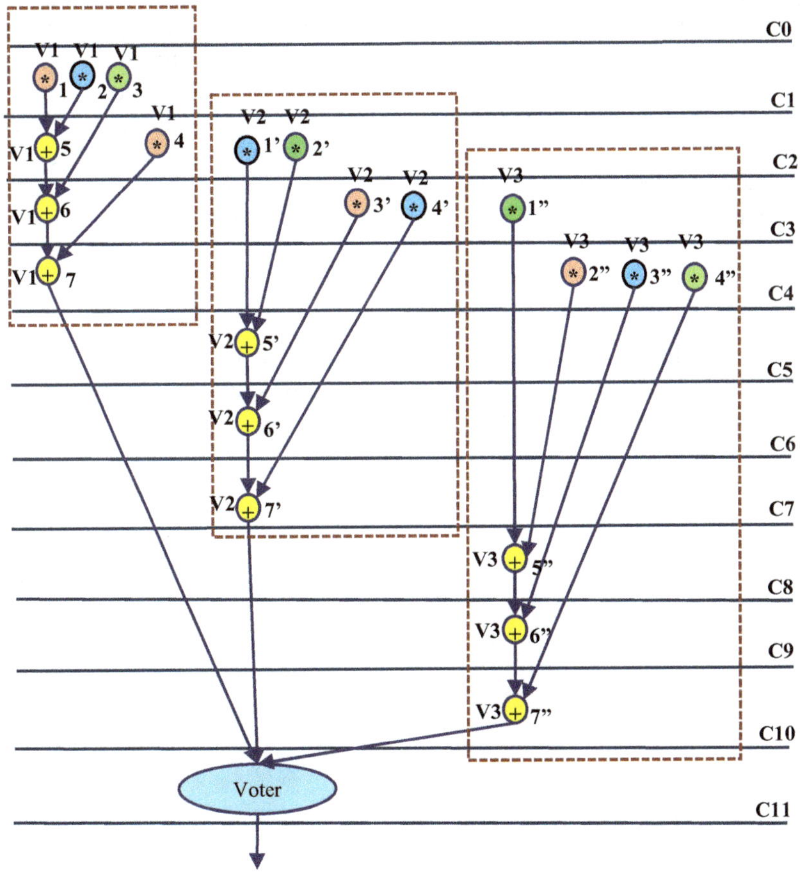

Figure 8.5 Scheduled data flow graph (SDFG) (TMR) of 4-point DCT with three multipliers (*) and one adder (+)

integration of a distinct multi-vendor allocation policy enhances the design's resilience against supply chain threats, ensuring that any single malicious insertion does not compromise overall system functionality.

8.10 Analysis of case studies

This section presents the analytical evaluation of the *TMR-based Trojan-resistant framework* across different hardware IP benchmarks. Each benchmark was executed for ten independent runs, and the average values were recorded to ensure statistical reliability (Sengupta *et al.*, 2023). During design-space exploration, equal weights were assigned to area and latency objectives ($\alpha_1 = \alpha_2 = 0.5$) in the normalized design-cost function to ensure a balanced trade-off between hardware area and scheduling delay. This equal weighting prevents bias toward either compactness or speed and reflects a realistic optimization objective for IP designers, who must simultaneously manage cost, latency, and Trojan-resilience constraints. The PSO-DSE parameters were adopted from Mishra and Sengupta (2014), *viz.*, acceleration coefficients $c_1 = c_2 = 2$; inertia weight linearly decreased from 0.9 to 0.1; swarm sizes of 3, 5, and 7; and random coefficients $r_1 = r_2 = 1$. Benchmarks included a variety of DSP cores, such as discrete cosine transform (DCT), finite impulse response (FIR), adaptive recursive filter (ARF), discrete wavelet transform (DWT), and JPEG, chosen to demonstrate scalability from small to large designs (Sengupta *et al.*, 2023; UCSB, 2025; Jain and Panda, 2007).

8.10.1 Design cost and convergence analysis

The PSO module integrated with the TMR scheduling framework efficiently generates low-cost (optimal) Trojan-resistant resource configuration. Figures 8.6 and 8.7 illustrate convergence and exploration time comparison for different swarm sizes.

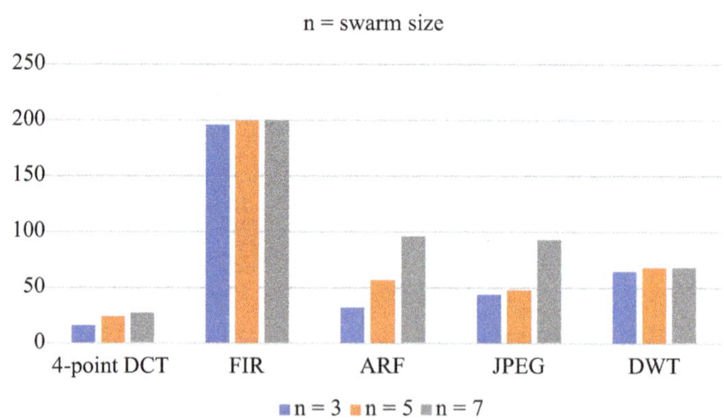

Figure 8.6 Convergence time comparison for generating optimal Trojan-resistant architecture w.r.t. different swarm sizes (n)

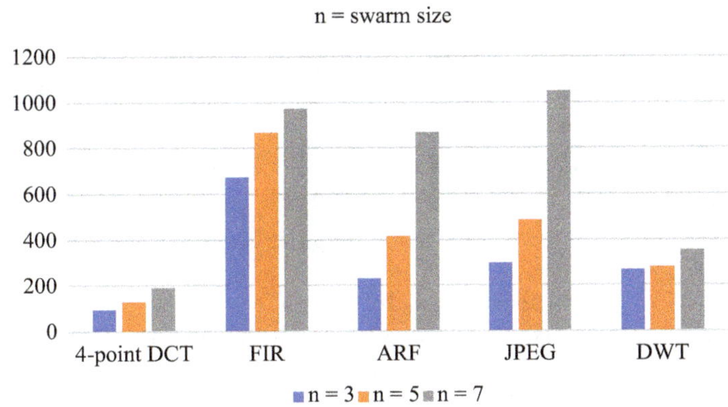

Figure 8.7 Exploration time comparison for generating optimal Trojan-resistant architecture w.r.t. different swarm sizes (n)

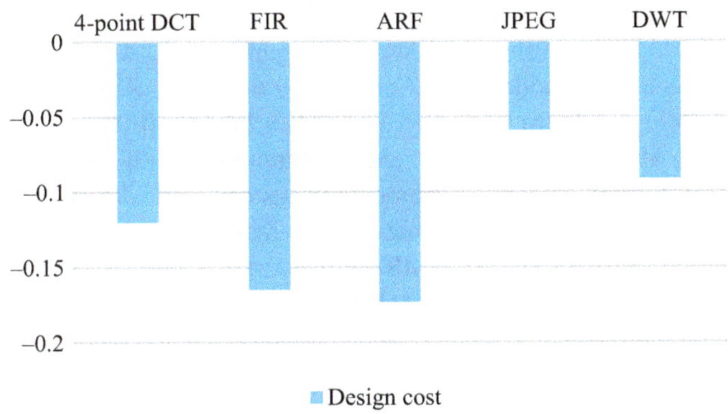

Figure 8.8 TMR design cost corresponding to different benchmarks

Convergence analysis showed that the time required to reach the optimal solution increased modestly with swarm size due to the larger number of particles evaluated per iteration, yet the final solution remained identical across all swarm sizes. This behavior validates the exploration capability of PSO-DSE to locate the global optimum even with limited swarm populations. Next, Figure 8.8 highlights the optimal resource configuration and the Trojan-resistant IP design cost corresponding to the TMR design. The optimal architecture solution obtained for the evaluated DSP benchmarks consistently involved minimal functional resources while maintaining complete Trojan-resistant/isolation capability. For instance, the 4-point DCT and FIR IP cores converged to configurations using three adders and nine multipliers distributed across three distinct micro-3PIP vendors.

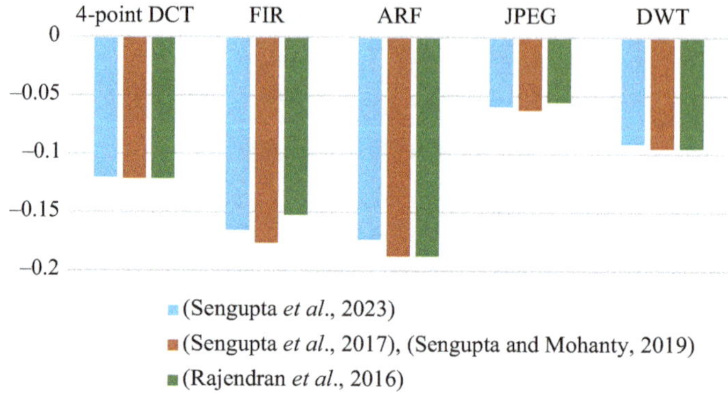

Figure 8.9 Design cost comparison of Sengupta et al. (2023) with Sengupta et al. (2017), Sengupta and Mohanty (2019), and Rajendran et al. (2016)

8.10.2 Security evaluation and comparative assessment

The TMR-based Trojan-resistant framework (Sengupta *et al.*, 2023) is compared with other Trojan detection methodologies (Sengupta *et al.*, 2017; Sengupta and Mohanty, 2019; Rajendran *et al.*, 2016) (that rely on DMR scheduling). While Sengupta *et al.* (2017), Sengupta and Mohanty (2019), and Rajendran *et al.* (2016) primarily identify the presence of Trojans, the TMR framework ensures both detection and isolation through its distinct multi-vendor allocation and majority-voting policy. Figure 8.9 highlights the comparison of the TMR framework (Sengupta *et al.*, 2023) with Sengupta *et al.* (2017), Sengupta and Mohanty (2019), and Rajendran *et al.* (2016).

8.10.3 Vulnerability analysis

A detailed vulnerability analysis has been conducted to evaluate how effectively the framework neutralizes untrusted micro-3PIP risks. The analysis depicts that the discussed TMR approach successfully mitigates hundreds of potential vulnerabilities across DSP design frameworks such as DCT, FIR, JPEG, DWT, etc. Table 8.1 illustrates the potential vulnerabilities handled by the TMR methodology (Sengupta *et al.*, 2023), highlighting its scalability from small arithmetic kernels to complex multimedia architectures.

8.10.4 Optimality validation of the proposed framework

To quantitatively validate the optimization quality of PSO-DSE within the TMR framework, multiple Pareto-front evaluation metrics have been analyzed, such as *generational distance* (GEN), *spacing* (SPA), *spread* (SPD), and *weighted sum* (WES). The optimality analysis indicates either zero or near-zero values for both GEN and SPA, confirming that the explored solutions lie very close to the true Pareto front and exhibit uniform distribution along it. Subsequently, low spreading and the weighted-sum metric further demonstrate comprehensive coverage of extreme design

Table 8.1 Potential vulnerabilities handled by TMR Trojan-resistant methodology

Benchmarks	**Total # of Vulnerabilities corresponding to DSP frameworks**
4-DCT	21
8-DCT	45
FIR	69
ARF	84
JPEG	99
MESA	273
IDCT	126
WDF	102
DWT	51

trade-offs across area and latency objectives. Collectively, these results substantiate that the PSO-assisted TMR Trojan-resistant framework (Sengupta *et al.*, 2023) achieves an *optimal balance among cost, performance metric, and security*, providing a robust, scalable, and computationally efficient solution for hardware IP design. The details of the analysis are available at Sengupta *et al.* (2023).

8.10.5 Advantages and limitations of low-cost TMR methodology (Sengupta et al.*, 2023)*

Advantages:

- The TMR framework delivers a fully Trojan-resistant hardware IP architecture that ensures complete security against functional Trojans (that alter computational output). It autonomously detects, isolates, and neutralizes any malicious logic embedded within the system.
- Moreover, the approach outlines an integration strategy that combines the Trojan-resistant IP design with a PSO-DSE mechanism. By applying PSO, the design space is intelligently explored, allowing the selection of an optimal hardware Trojan-resistant configuration with minimal cost overhead.

Limitations:

- To mitigate the resource overhead resulting from the replication of functional units in the TMR-based design, carefully encoding the PSO-DSE framework becomes essential.
- The efficiency of PSO-DSE strongly depends on the careful tuning of its control parameters, such as inertia weight, random coefficients, and social and cognitive factors, to ensure timely and stable convergence toward the optimal solution.
- The TMR methodology assumes the availability of reliable fault-tolerant components, including a trusted voter and error detection circuits. These modules must be secure and Trojan-free in order to uphold the overall data integrity of the system.

8.11 Conclusion

This chapter presented a comprehensive framework for developing Trojan-resistant IP designs and effective Trojan isolation strategies for secure IP designs. The discussed chapter integrates design space exploration with a redundancy-based security framework to ensure robust defense against functional hardware Trojans with optimal design architecture. By intelligently managing design trade-offs and multi-vendor hardware allocations, the approach achieves high reliability and resilience with minimal cost overhead.

8.12 Questions and exercises

1. What is the primary difference between application-specific computing systems and general-purpose computing systems?
2. Why is it important to design application-specific computing cores? Explain the high-level synthesis process in brief and its importance in designing application-specific computing cores.
3. Discuss some of the real-world data and computation-intensive applications and their applications.
4. Explain functional Trojans and their effects.
5. Why is it difficult to detect the presence of functional Trojans in the hardware IP design using conventional Trojan detection mechanisms?
6. Discuss the need for designing Trojan-resistant hardware IP.
7. Why is it important to generate an optimal design in the context of Trojan-resistant hardware architecture?
8. Discuss some state-of-the-art in the field of hardware Trojan detection.
9. Explain the threat model (considered in this chapter) in detail.
10. Discuss some real-world threat scenarios for functional hardware Trojans.
11. Explain the TMR architecture and its importance in detail.
12. Explain the distinct multi-vendor allocation policy and its importance in detail.
13. How the TMR Trojan-resistant framework prevents collusion between multiple vendors for the same Trojan payload.
14. Draw an overview diagram for the optimal Trojan-resistant methodology and discuss each step in detail.
15. What are the advantages of PSO-DSE, and also discuss the importance of the integration of PSO-DSE with the TMR Trojan-resistant framework?
16. Explain the details of the PSO-DSE process with the help of a flow chart.
17. Demonstrate the Trojan-resistant methodology using real-world data-intensive applications, such as FIR, DWT, etc.
18. What are the advantages and limitations of the low-cost Trojan-resistant methodology?

References

C. Pilato, S. Garg, K. Wu, R. Karri, and F. Regazzoni, (2018), "Securing hardware accelerators: a new challenge for high-level synthesis," *IEEE Embedded Systems Letters*, vol. 10, no. 3, pp. 77–80, 2018.

A. Anshul, and A. Sengupta, (2023a), "A survey of high level synthesis based hardware security approaches for reusable IP cores," *IEEE Circuits and Systems Magazine*, vol. 23, no. 4, pp. 44–62, Fourthquarter 2023.

B. C. Schafer, and Z. Wang, (2020) "High-level synthesis design space exploration: past, present, and future," *IEEE Transactions on Computer-Aided Design of Integrated Circuits and Systems*, vol. 39, no. 10, pp. 2628–2639, 2020.

B. K. Mohanty, and P. K. Meher, (2016), "A high-performance FIR filter architecture for fixed and reconfigurable applications," *IEEE Transactions on Very Large Scale Integration (VLSI) Systems*, vol. 24, no. 2, pp. 444–452, 2016.

M. Rathor, A. Anshul, and A. Sengupta, (2024), "Securing reusable IP cores using voice biometric based watermark," *IEEE Transactions on Dependable and Secure Computing*, vol. 21, no. 4, pp. 2735–2749, 2024.

M. Rostami, F. Koushanfar, and R. Karri, (2014), "A primer on hardware security: models, methods, and metrics," *Proceedings of the IEEE*, vol. 102, no. 8, pp. 1283–1295, 2014.

A. Sengupta, A. Anshul, V. Chourasia, and N. Kumar, (2024), "M-HLS: malevolent high-level synthesis for watermarked hardware IPs," *IEEE Embedded Systems Letters*, vol. 16, no. 4, pp. 497–500, 2024.

A. Anshul, and A. Sengupta, (2023b), "Exploration of optimal crypto-chain signature embedded secure JPEG-CODEC hardware IP during high level synthesis," *Microprocessors and Microsystems*, vol. 102, p. 104916, 2023.

A. Sengupta, and M. Rathor, (2021), "Facial biometric for securing hardware accelerators," *IEEE Transactions on Very Large Scale Integration (VLSI) Systems*, vol. 29, no. 1, pp. 112–123, 2021.

A. Sengupta, and M. Rathor, (2020), "Securing hardware accelerators for CE systems using biometric fingerprinting," *IEEE Transactions on Very Large Scale Integration (VLSI) Systems*, vol. 28, no. 9, pp. 1979–1992, 2020.

R. Chaurasia, A. Anshul, A. Sengupta, and S. Gupta, (2022), "Palmprint biometric versus encrypted hash based digital signature for securing DSP cores used in CE systems," *IEEE Consumer Electronics Magazine*, vol. 11, no. 5, pp. 73–80, 2022.

A. Sengupta, (2017a), "Hardware vulnerabilities and its effect on CE devices: design-for-security against Trojan," *IEEE Consumer Electronics Magazine*, vol. 6, no. 3, pp. 126–133, 2017.

A. Sengupta, (2017b), "Hardware security of CE devices: threat models and defence against IP Trojans and IP piracy," *IEEE Consumer Electronics Magazine*, vol. 6, no. 1, pp. 130–134, 2017.

M. Fyrbiak *et al.*, (2019), "HAL—the missing piece of the puzzle for hardware reverse engineering, Trojan detection and insertion," *IEEE Transactions on Dependable and Secure Computing*, vol. 16, no. 3, pp. 498–510, 2019.

A. Sengupta, A. Anshul, and R. Chaurasia, (2023), "Exploration of optimal functional Trojan-resistant hardware intellectual property (IP) core designs during high level synthesis," *Microprocessors and Microsystems*, vol. 103, p. 104973, 2023.

R. Schneiderman, (2010), "DSPs evolving in consumer electronics ap plications," *IEEE Signal Processing Magazine*, vol. 27, no. 3, pp. 6–10, May 2010.

A. Sengupta, A. Anshul, V. Chourasia, and N. Bhui, (2025), "Security vulnerability (backdoor Trojan) during machine learning accelerator design phases," *IT Professional*, vol. 27, no. 1, pp. 65–72, 2025.

A. Sengupta, S. Bhadauria, and S. P. Mohanty, (2017), "TL-HLS: methodology for low cost hardware Trojan security aware scheduling with optimal loop unrolling factor during high level synthesis," *IEEE Transactions on Computer-Aided Design of Integrated Circuits and Systems*, vol. 36, no. 4, pp. 655–668, 2017.

A. Sengupta, and S. Mohanty, (2019), "Trojan security aware DSP IP core and integrated circuits," *IET Book: IP Core Protection and Hardware-Assisted Security for Consumer Electronics*, 2019, doi:10.1049/PBCS060E_ch3.

Liao, Y, Adegbija, T, and Lysecky, R. (2023), "Efficient system-level design space exploration for high-level synthesis using pareto-optimal subspace pruning," *2023 28th Asia and South Pacific Design Automation Conference (ASP-DAC)*, Tokyo, Japan, pp. 1–6.

Qian, X, Shi, J, Shi, L, Zhang, H, Bian, L, and Qian, W., (2022), "Scheduling information-guided efficient high-level synthesis design space exploration," *2022 IEEE 40th International Conference on Computer Design (ICCD)*, Olympic Valley, CA, USA, pp. 203–206.

Liu, H.-Y., and Carloni, L. P., (2013), "On learning-based methods for design-space exploration with High-Level Synthesis," *2013 50th ACM/EDAC/IEEE Design Automation Conference (DAC)*, Austin, TX, 2013, pp. 1–7.

Williams, A. C., Brown, A. D., and Zwolinski, M., (2008), "Simultaneous optimisation of dynamic power, area and delay in behavioural synthesis," *Proceedings of IEEE Computers and Digital Techniques*, vol. 147, no. 6, pp. 383–390.

A. Sengupta, and S. Bhadauria, (2015), "Bacterial foraging driven exploration of multi cycle fault tolerant datapath based on power-performance tradeoff in high level synthesis," *Expert Systems with Applications*, vol. 42, no. 10, pp. 4719–4732, 2015.

A. J. Hu, (1997), "Formal hardware verification with BDDs: an introduction," *1997 IEEE Pacific Rim Conf. on Communications, Computers and Signal Processing, PACRIM. 10 Years Networking the Pacific Rim*, Victoria, BC, Canada, 1987–1997, pp. 677–682, vol. 2, 1997.

M. Bushnell, and V. Agrawal, (2001), "Essentials of electronic testing for digital, memory, and mixed-signal VLSI circuits," *IEEE Circuits and Devices Magazine*, vol. 17, no. 4, pp. 39–40, 2001.

M. Tehranipoor, and F. Koushanfar, (2010), "A survey of hardware trojan taxonomy and detection," *IEEE Design & Test of Computers*, vol. 27, no. 1, pp. 10–25, 2010.

D. Agrawal, S. Baktir, D. Karakoyunlu, P. Rohatgi, and B. Sunar, (2007), "Trojan detection using IC fingerprinting," *2007 IEEE Symposium on Security and Privacy (SP '07)*, Berkeley, CA, USA, pp. 296–310, 2007.

S. Narasimhan, D. Du, R. Subhra Chakraborty, S. Paul, F. Wolff1, C. Papachristou *et al.*, (2010), "Multiple-parameter side-channel analysis: A non-invasive hardware Trojan detection approach," *2010 IEEE International Symposium on Hardware-Oriented Security and Trust (HOST)*, Anaheim, CA, USA, 2010, pp. 13–18, doi:10.1109/HST.2010.5513122.

X. Wang, H. Salmani, M. Tehranipoor, and Jim Plusquellic, (2008), "Hardware Trojan Detection and Isolation Using Current Integration and Localized Current Analysis," *2008 IEEE International Symposium on Defect and Fault Tolerance of VLSI Systems*, Cambridge, MA, USA, 2008, pp. 87–95, doi:10.1109/DFT.2008.61.

J. J. Rajendran, O. Sinanoglu, and R. Karri, (2016), "Building trustworthy systems using untrusted components: a high-level synthesis approach," *IEEE Transactions on Very Large Scale Integration (VLSI) Systems*, vol. 24, no. 9, pp. 2946–2959, 2016.

H. Martin, L. Entrena, S. Dupuis, and G. Di Natale, "A Novel Use of Approximate Circuits to Thwart Hardware Trojan Insertion and Provide Obfuscation," *2018 IEEE 24th International Symposium on On-Line Testing And Robust System Design (IOLTS)*, Platja d'Aro, Spain, pp. 41–42, doi:10.1109/IOLTS.2018.8474077.

M. Mehdinia Siboni, and S. Mohammadhossein Shekarian, "An Effective Application of Obfuscated TMR to Tackle Hardware Trojans," *2019 9th International Conference on Computer and Knowledge Engineering (ICCKE)*, Mashhad, Iran, 2019, pp. 12–17, doi:10.1109/ICCKE48569.2019.8964879.

N. B. Gunti, and K. Lingasubramanian, (2017), "Neutralization of the effect of hardware Trojan in SCADA system using selectively placed TMR," *2017 IEEE International Symposium on Nanoelectronic and Information Systems (iNIS)*, pp. 99–104, 2017.

H. Li, A. Abdelhadi, R. Shi, J. Zhang, and Q. Liu, (2021), "Adversarial hardware with functional and topological camouflage," *IEEE Transactions on Circuits and Systems II: Express Briefs*, vol. 68, no. 5, pp. 1685–1689, 2021.

M. Abderehman, R. Gupta, R. R. Theegala, and C. Karfa, (2022a), "BLAST: belling the black-hat high-level synthesis tool," *IEEE Transactions on Computer-Aided Design of Integrated Circuits and Systems*, vol. 41, no. 11, pp. 3661–3672, 2022.

A. Kulkarni, Y. Pino, and T. Mohsenin, (2016), "Adaptive real-time Trojan detection framework through machine learning," *2016 IEEE International Symposium on Hardware Oriented Security and Trust (HOST)*, McLean, VA, USA, pp. 120–123, 2016.

P. Fezzardi, F. Ferrandi, and C. Pilato, (2018), "Enabling automated bug detection for IP-based designs using high-level synthesis," *IEEE Design & Test*, vol. 35, no. 5, pp. 54–62, 2018.

M. Abderehman, T. Rakesh Reddy, and C. Karfa, (2022b), "DEEQ: data-driven end-to-end EQuivalence checking of high-level synthesis," *2022 23rd International Symposium on Quality Electronic Design (ISQED)*, Santa Clara, CA, USA, pp. 64–70, 2022.

N. Fern, and K.-T. Cheng, (2019), "Evaluating assertion set completeness to expose hardware Trojans and verification blindspots," *2019 Design, Automation & Test in Europe Conference & Exhibition (DATE)*, Florence, Italy, pp. 402–407, 2019.

U. Alsaiari, and F. Gebali, (2019), "Hardware Trojan detection using reconfigurable assertion checkers," *IEEE Transactions on Very Large Scale Integration (VLSI) Systems*, vol. 27, no. 7, pp. 1575–1586, 2019.

S. Bhunia, M. S. Hsiao, M. Banga, and S. Narasimhan, (2014), "Hardware Trojan attacks: threat analysis and countermeasures," *Proceedings of the IEEE*, vol. 102, no. 8, pp. 1229–1247, 2014.

W. Hu, C.-H. Chang, A. Sengupta, S. Bhunia, R. Kastner, and H. Li, (2021), "An overview of hardware security and trust: threats, countermeasures, and design tools," *IEEE Transactions on Computer-Aided Design of Integrated Circuits and Systems*, vol. 40, no. 6, pp. 1010–1038, 2021.

A. Sengupta, R. Chaurasia, and T. Reddy, (2021), "Contact-less palmprint biometric for securing DSP coprocessors used in CE systems," *IEEE Transactions on Consumer Electronics*, vol. 67, no. 3, pp. 202–213, 2021.

R. V. Kshirsagar, and R. M. Patrikar, (2009), "Design of a novel fault-tolerant voter circuit for TMR implementation to improve reliability in digital circuits," *Microelectronics Reliability*, vol. 49, no. 12, pp. 1573–1577, 2009.

D. Kachave, and A. Sengupta, (2016), "Integrating physical level design and high level synthesis for simultaneous multi-cycle transient and multiple transient fault resiliency of application specific datapath processors," *Elsevier Journal on Microelectronics Reliability*, vol. 60, pp. 141–152, 2016.

V. Krishnan, and S. Katkoori, (2006), "A genetic algorithm for the design space exploration of datapaths during high-level synthesis," *IEEE Transactions on Evolutionary Computation*, vol. 10, no. 3, pp. 213–229, 2006.

Kao, Y., Chen, M., and Huang, Y, (2012), "A hybrid algorithm based on ACO and PSO for capacitated vehicle routing problems," *Mathematical Problems in Engineering*, vol. 2012, Article ID 123456, doi:10.1155/2012/726564.

A. Sengupta, and S. Bhadauria, (2014), "Automated exploration of datapath in high level synthesis using temperature dependent bacterial foraging optimization algorithm," *2014 IEEE 27th Canadian Conference on Electrical and Computer Engineering (CCECE)*, Toronto, ON, Canada, pp. 1–5, 2014.

V. Mishra, and A. Sengupta, (2014), "MO-PSE: adaptive multi-objective particle swarm optimization based design space exploration in architectural synthesis for application specific processor design," *Advances in Engineering Software*, vol. 67, pp. 111–124, 2014.

UCSB, (2025), University of California Santa Barbara Express Group, accessed on Sept. 2025. [Online]. Available: http://express.ece.ucsb.edu/benchmark/

R. Jain, and P. R. Panda, (2007), "An efficient pipelined VLSI architecture for lifting based 2D-discrete wavelet transform," *Proceedings of the IEEE International Symposium on Circuits and Systems (ISCAS)*, New Orleans, LA, USA, pp. 1377–1380, May 2007.

Part IV

Additional perspectives

Chapter 9

Compromising IP designs via malicious exploitation of commercial CAD-HLS tools vs. backdoor Trojan-inserted HLS frameworks

Anirban Sengupta[1]

The chapter provides the readers with an introduction to the computer-aided design (CAD) tools. These CAD tools are an integral part of computing system design. Owing to the time-to-market demand augmented with the expectations of high design productivity, third-party CAD-based high-level synthesis (3P-HLS) tools have become extremely popular. However, these 3P-HLS tools are often considered untrustworthy, as they have been shown to comprise malicious logic in the form of a backdoor Trojan (furtively inserted by a rogue tool developer). This chapter shows how reusable intellectual property (IP) designs can be covertly compromised using two techniques (attack vectors), *viz.*, (a) malicious exploitation of a commercial CAD-HLS tool for latency-degradation Trojan payload and (b) backdoor Trojan insertion within the HLS framework/design flow. The chapter also provides a comparative perspective on the Trojan payload impact of both attacking techniques as well as their associated area/resource overheads. Finally, discuss the evasive capability of the HLS Trojan attacks against known detection techniques.

The rest of the chapter is structured as follows: Section 9.1 provides the introduction of the chapter; Section 9.2 outlines the state-of-the-art; Section 9.3 presents a brief discussion on the Trojan insertion technique; Section 9.4 presents an analysis of case studies; Section 9.5 concludes the chapter by summarizing its significance and implications.

9.1 Introduction

Another alarming feature of such HLS Trojans is that they are stealthy and elusive by nature, which means they can outmaneuver known detection techniques. These HLS Trojans can have multiple types of payloads, *viz.*, functional output change, latency/performance degradation, leakage of confidential information, denial of service, and data damage. Additionally, another facet of these Trojans is that they

[1]Department of Computer Science and Engineering, Indian Institute of Technology Indore, India

can be furtively injected into any phase of the HLS-based IP design process, which makes the detection process extremely convoluted (Rathor and Sengupta, 2024; Sengupta *et al.*, 2024). In this chapter, we show how IP designs can be covertly compromised using two techniques, *viz.*, (a) malicious exploitation of a commercial CAD-HLS tool for a latency-degradation Trojan payload, (b) backdoor Trojan insertion within the HLS framework/design flow. The chapter also compares the Trojan payload impact of both attacking techniques as well as their associated area/resource overheads.

9.2 State-of-the-art

In the domain of HLS hardware Trojan detection, various techniques have been explained in the state-of-the-art, such as Cui *et al.* (2014), Veeranna and Schafer (2017), Rajmohan *et al.* (2020), Rathor and Sengupta (2024), Sengupta *et al.* (2017), Sengupta *et al.* (2023), Yasaei *et al.* (2022), Abderehman *et al.* (2022), and Jin and Makris (2008), to detect, resist, and isolate hardware Trojans. Cui *et al.* (2014) highlighted design system rules that facilitate run-time Trojan detection and fast recovery by leveraging third-party IP core diversity. However, Cui *et al.* (2014) target only Trojans affecting logic values and lack the capability to detect Trojans that introduce performance or latency degradation. Similarly, Veeranna and Schafer (2017) employed formal verification to detect functional Trojans such as IP malfunction, information leakage, and denial of service in third-party behavioral IPs, but did not cover performance or latency-degradation attacks. Next, Rajmohan *et al.* (2020) have discussed a particle swarm optimization-based design space exploration (PSO-DSE) technique combined with a Trojan detection mechanism that minimizes resource usage. This methodology employed redundancy logic using at least three distinct IP vendors, yet it remains limited to detecting functional Trojans only. Subsequently, Rathor and Sengupta (2024) discussed a high-level transformation (HLT)-based Trojan detection methodology focused on identifying battery exhaustion-based Trojans. Next, the approach in Sengupta *et al.* (2017) focused on constraint-driven, low-cost dual modular redundancy (DMR)-based scheduling in HLS to detect functional Trojans. However, Rathor and Sengupta (2024) and Sengupta *et al.* (2017) failed to detect performance or latency-degrading Trojans. Sengupta *et al.* (2023), a PSO-DSE methodology-driven triple modular redundancy (TMR) framework with multi-vendor strategies providing complete detection and isolation of only functional hardware Trojans. Other methodologies, such as Yasaei *et al.* (2022), Abderehman *et al.* (2022), and Jin and Makris (2008), rely on Pyverilog, C-to-RTL equivalence checking, and path delay fingerprinting, respectively, to detect hardware Trojans. However, Yasaei *et al.* (2022) and Jin and Makris (2008) are not scalable for complex HLS-based IP designs due to the weaker learning behavior of Pyverilog and the exponential increase in the number of delay paths and test patterns, respectively. Additionally, Abderehman *et al.* (2022) failed to detect latency and performance degradation Trojans. Moreover, none of these state-of-the-arts (Cui *et al.*, 2014; Veeranna and Schafer, 2017;

Rajmohan *et al.*, 2020; Rathor and Sengupta, 2024; Sengupta *et al.*, 2017, 2023; Yasaei *et al.*, 2022; Abderehman *et al.*, 2022; Jin and Makris, 2008) evaluated or compared hardware Trojan impacts across commercial CAD tools (such as Xilinx Vivado HLS) (AMD Vivado™ Design Suite, 2025) and academic HLS tools (CAD for Assurance, 2025) as well as are not capable of detecting latency/performance degradation-based HLS-aided Trojan. In contrast, this chapter discusses how a commercial CAD-HLS tool can be maliciously exploited for inducing latency-degradation Trojan payloads *vs.* how a backdoor Trojan can be covertly inserted in an HLS framework/flow by a rouge HLS tool developer.

9.3 Discussion on Trojan insertion technique

9.3.1 Malicious exploitation of a commercial CAD-HLS tool for latency-degradation Trojan payload

Figure 9.1 shows the Trojan attack in the Xilinx Vivado HLS tool flow. The Vivado HLS design flow simplifies hardware design by allowing developers to use C, C++, or System C as input to describe functionality. The process begins with C functional verification, ensuring that the input code behaves as intended through simulation. Next, the tool performs high-level synthesis, converting the verified C-code into register-transfer level (RTL) code (typically in VHDL or Verilog). Following this, C/RTL co-simulation is carried out to validate that the synthesized RTL behaves identically to the original C model. Moreover, to optimize performance, area, or latency, designers apply directives and constraints, guiding the synthesis tool in areas like loop unrolling, pipelining, etc. After applying these

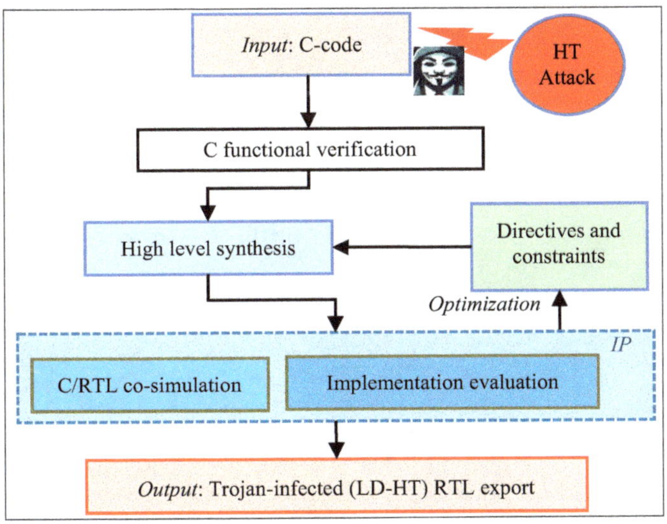

Figure 9.1 Xilinx Vivado HLS flow with potential Trojan attack

optimizations, the design undergoes implementation evaluation to assess metrics such as timing, resource utilization, and throughput. Finally, the design is exported as RTL, ready for downstream FPGA implementation using Vivado's synthesis, placement, and routing tools.

The primary target of the attacker (i.e., rouge IP designer) is to maliciously exploit the commercial CAD tool (such as Xilinx Vivado HLS) to generate a Trojan-infected (compromised) IP design, which is sent to the system-on-chip (SoC) design house for further integration into the final integrated circuit (IC). As shown in Figure 9.1, an attacker can launch a latency-degradation hardware Trojan (LD-HT) attack by manipulating the input C-code of the application. Figure 9.2 shows an example, the partial view of Trojan-inserted C-code of a finite impulse response (FIR) filter. The Trojan trigger and payload are shown in red. As shown in Figure 9.2, the FIR C-code executes normal iterations on input values to produce the final output in normal conditions. However, when the Trojan is triggered (i.e., z == x[i]), the FIR C-code executes dummy computations under two additional iterations, causing latency overhead. This is specifically due to additional read operations performed for executing the dummy computations. Next, Figure 9.3 illustrates the equivalent CDFG of Trojan-infected FIR filter C-code (shown in Figure 9.2. The operation shown in black color represents the normal computations of the FIR filter; however, the operations shown in red are the dummy computations inserted by an attacker to enable latency degradation without affecting the functional output. Further, the LD-HT activates/triggers only under an attacker's defined, specific, and rare condition. As this Trojan does not affect the functional output computation and remains inactive under normal conditions, it is therefore very challenging to detect (and stealthy/elusive by nature). Next, Figure 9.4 shows the partial view of the VHDL code generated using Vivado HLS (post C-to-RTL synthesis) corresponding to Trojan-free and Trojan-infected FIR C-codes.

```
int x = [], int Y = [], int h = [], int z, int i;
if (z != x[i])          <-- Trojan trigger
{
for (int i = 0; i < 8; i++)
{
Y[i+1] = Y[i] + (h[i]*x[n – i]);
}
}
else
{
for (int i=0; i<10; i++)          <-- Trojan payload
{
Y[i+1] =Y[i] + (h[i]*x[n – i]);
}
}
```

Figure 9.2 Partial view of Trojan-inserted C-code of a finite impulse response (FIR) Filter

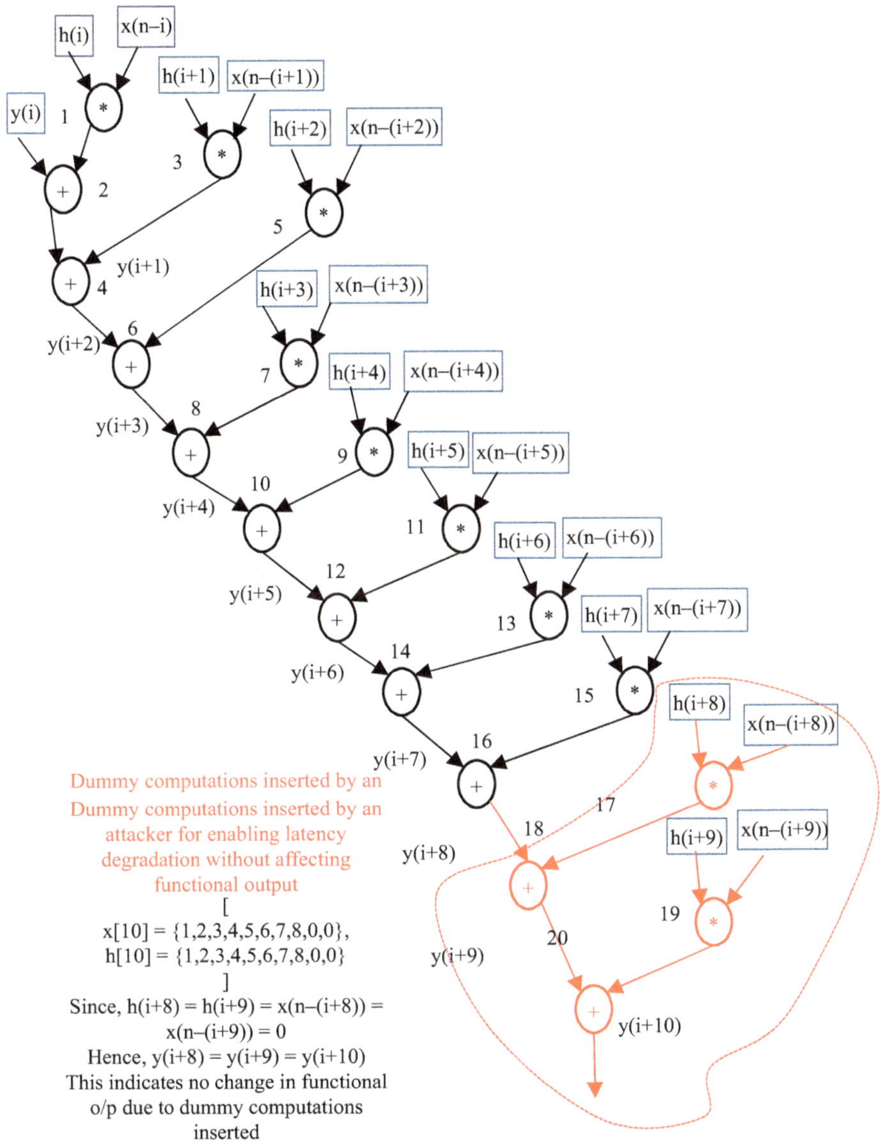

Figure 9.3 Equivalent CDFG of Trojan-infected FIR filter C-code

The generated VHDL representations of both Trojan-free and Trojan-infected FIR C-code are similar in terms of complexity and size. This indicates the difficulty faced by an SoC designer in tracing the malicious Trojan logic and is therefore very cumbersome to detect. A very close controlled inspection of the two VHDL codes (Trojan-free—which is the golden model and the Trojan-infected one) reveals some subtle changes, which a rouge IP designer can only know. It is well known that an IP designer only provides the IP core to the SoC integrator, without providing any golden model (Sengupta *et al.*, 2017). It is therefore impossible for an SoC

```
architecture behav of main is
attribute CORE_GENERATION_INFO : STRING;
attribute CORE_GENERATION_INFO of behav : architecture is
"main,hls_ip_2014_3,{HLS_INPUT_TYPE=cxx,HLS_INPUT_FLOAT=0,HL
S_INPUT_FIXED=0,HLS_INPUT_PART=xc7vx690tffg1761-
2,HLS_INPUT_CLOCK=10.000000,HLS_INPUT_ARCH=others,HLS_SYN_
CLOCK=8.680000,HLS_SYN_LAT=25,HLS_SYN_TPT=none,HLS_SYN_M
EM=0,HLS_SYN_DSP=1,HLS_SYN_FF=85,HLS_SYN_LUT=156}";
constant ap_const_logic_1 : STD_LOGIC := '1';
constant ap_const_logic_0 : STD_LOGIC := '0';
constant ap_ST_st1_fsm_0 : STD_LOGIC_VECTOR (1 downto 0) := "00";
constant ap_ST_st2_fsm_1 : STD_LOGIC_VECTOR (1 downto 0) := "01";
constant ap_ST_st3_fsm_2 : STD_LOGIC_VECTOR (1 downto 0) := "10";
constant ap_ST_st4_fsm_3 : STD_LOGIC_VECTOR (1 downto 0) := "11";
constant ap_const_lv1_0 : STD_LOGIC_VECTOR (0 downto 0) := "0";
constant ap_const_lv3_0 : STD_LOGIC_VECTOR (2 downto 0) := "000";
constant ap_const_lv4_0 : STD_LOGIC_VECTOR (3 downto 0) := "0000";
constant ap_const_lv32_0 : STD_LOGIC_VECTOR (31 downto 0) :=
```

Trojan-free VHDL code

```
architecture behav of main is
attribute CORE_GENERATION_INFO : STRING;
attribute CORE_GENERATION_INFO of behav : architecture is
"main,hls_ip_2014_3,{HLS_INPUT_TYPE=cxx,HLS_INPUT_FLOAT=0,HL
S_INPUT_FIXED=0,HLS_INPUT_PART=xc7vx690tffg1761-
2,HLS_INPUT_CLOCK=10.000000,HLS_INPUT_ARCH=others,HLS_SYN_
CLOCK=8.680000,HLS_SYN_LAT=27,HLS_SYN_TPT=none,HLS_SYN_M
EM=0,HLS_SYN_DSP=1,HLS_SYN_FF=86,HLS_SYN_LUT=171}";
constant ap_const_logic_1 : STD_LOGIC := '1';
constant ap_const_logic_0 : STD_LOGIC := '0';
constant ap_ST_st1_fsm_0 : STD_LOGIC_VECTOR (1 downto 0) := "00";
constant ap_ST_st2_fsm_1 : STD_LOGIC_VECTOR (1 downto 0) := "01";
constant ap_ST_st3_fsm_2 : STD_LOGIC_VECTOR (1 downto 0) := "10";
constant ap_ST_st4_fsm_3 : STD_LOGIC_VECTOR (1 downto 0) := "11";
constant ap_const_lv1_0 : STD_LOGIC_VECTOR (0 downto 0) := "0";
constant ap_const_lv4_0 : STD_LOGIC_VECTOR (3 downto 0) := "0000";
constant ap_const_lv32_0 : STD_LOGIC_VECTOR (31 downto 0) :=
"00000000000000000000000000000000";
```

Trojan-infected VHDL code

Figure 9.4 Partial view of register-transfer level (RTL) (VHDL code) corresponding to synthesized Trojan-free and Trojan-infected FIR C-code

integrator, who integrates numerous pre-verified third-party IP designs, to detect subtle Trojan logic deeply hidden inside the complex RTL codes before integration (Sengupta *et al.*, 2017, 2024). For example, some minor changes in the Trojan VHDL code are in terms of signal width, number of internal signals, and number of flip-flop (FF)/look-up tables (LUTs), which the specific IP vendor can only know.

9.3.2 Backdoor Trojan-inserted HLS frameworks

Figure 9.5 shows the backdoor Trojan (performance degradation hardware Trojan (PD-HT)) insertion in an HLS design flow. As shown in Figure 9.5, the primary inputs comprise (a) a transfer function/intermediate representation of the application (CDFG), (b) resource configuration (i.e., adder, multipliers, etc.), which is equivalent to the pragma directive of commercial HLS tools, and (c) a module library corresponding to different functional components/resources. The HLS framework accepts these primary inputs and performs scheduling, allocation, and binding (key steps of the HLS framework). Subsequently, the mux-based interconnect design information is generated using the corresponding multiplexing scheme. As discussed in Sengupta *et al.* (2024), HLS-generated IP designs expose vulnerabilities that are exploited by an attacker for secret hardware Trojan insertion. The target of an attacker is to insert the Trojan in a safe location by leveraging the unutilized ports in the mux-interconnect design information of registers. Post PD-HT insertion, the final Trojan-inserted RTL datapath and controller design is generated.

A PD-HT refers to a covert and malicious alteration introduced into an IP design. Unlike conventional Trojans that disrupt system functionality, PD-HT is designed to subtly reduce performance (by adding unwarranted, larger delays). The delay is introduced by a series of an even number of inverters (v) added by an attacker in the mux-based interconnect design stage of the HLS process, where the extent of the performance degradation is manipulated by adjusting the number of inverters.

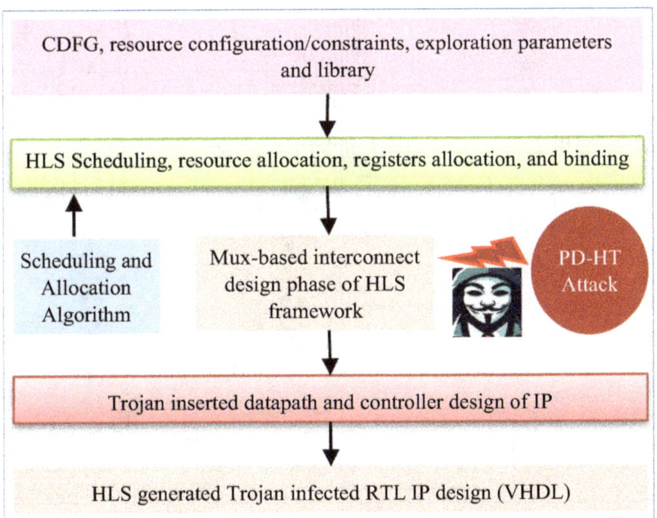

Figure 9.5 Backdoor Trojan (PD-HT) insertion in HLS flow

An attacker inserts PD-HT logic during the HLS tool development process. For example, Sengupta *et al.* (2024) have implemented PD-HT insertion in an academic/publicly available HLS tool (CAD for Assurance, 2025). Similarly, a rouge designer in any HLS tool development house can insert PD-HT during its design process. Figure 9.6 highlights the HLS framework-generated Trojan-infected RTL datapath (with PD-HT) corresponding to the FIR IP. The PD-HT is activated only when a specific condition (trigger) defined by the attacker is met. As shown in Figure 9.6, the trigger condition is monitored using the comparator circuitry (*W*), which triggers the Trojan logic when both inputs of the comparator are equal. When one of the

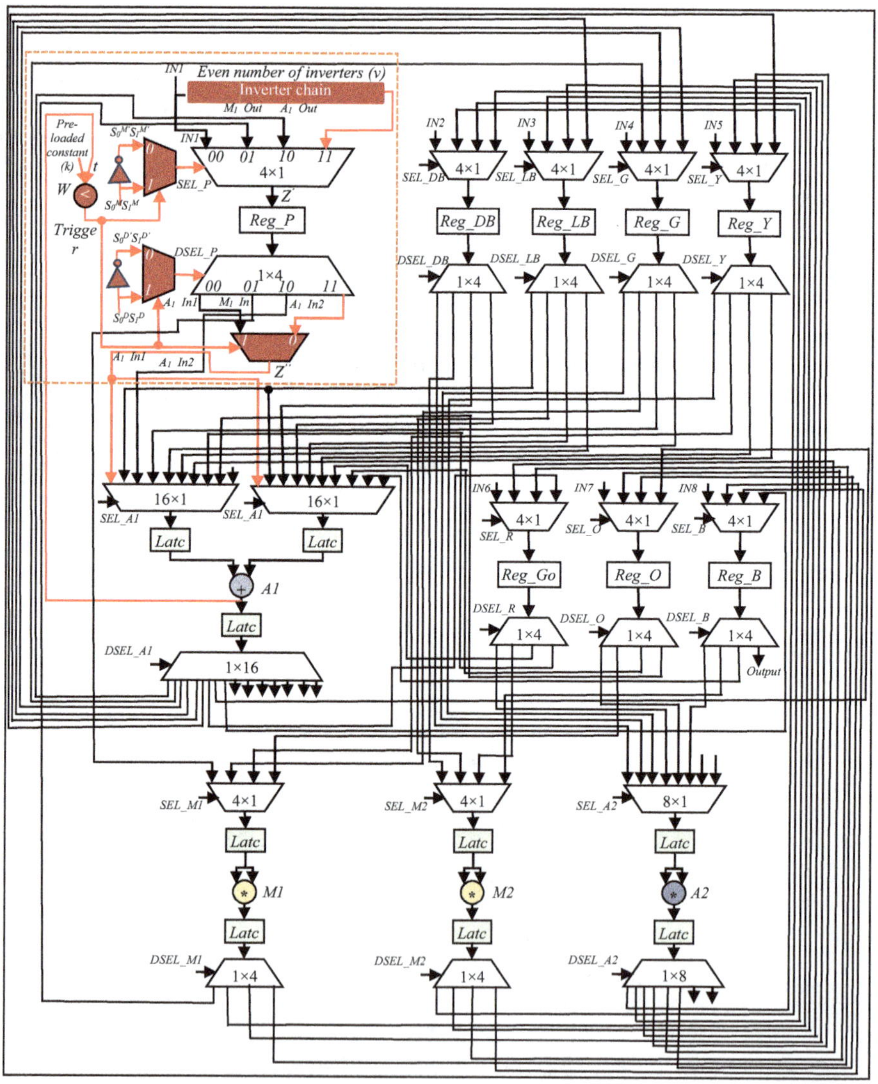

Figure 9.6 HLS framework-generated Trojan-infected RTL datapath (with PD-HT) corresponding to FIR IP design

comparator input signals (denoted as t) matches a constant value k (predetermined by the attacker), the Trojan logic is triggered. The signal t originates internally from the output of the functional unit (adder unit). Once the triggering condition is met ($t = k$), the circuit's output is intentionally delayed, impacting performance. If the condition is not met, the output behaves as expected, showing no signs of malice. As this Trojan does not affect the functional output computation and remains inactive under normal conditions, it is therefore very challenging to detect (and stealthy/elusive by nature). Further, Table 9.1 illustrates the behavior of the multiplexer linked to *Reg_P* (shown in Figure 9.6). When the condition $t = k$ is satisfied (indicating the activation of the Trojan), the output *Z''* reflects a delayed version of the IN1 signal, resulting in reduced performance. In contrast, when t does not equal k, *Z''* directly outputs the signal from *IN1* (specifically *A1_In1*), maintaining the circuit's expected performance without introducing any delay. Figure 9.7 shows the partial view of the VHDL code

Table 9.1 Behavioral table corresponding to Mux of REG_P in Figure 9.6) of PD-HT logic

Rare-event trigger	Comparator o/p (W)	S_o^M	S_1^M	F'	State
$k = t$ (active)	0	1	1	IN1	Trojan payload
$k \neq t$ (dormant)	1	0	0	IN1	Normal behavior
	1	0	1	M_{1Out}	
	1	1	0	A_{1Out}	

```
begin
---------------------------TROJAN BLOCK--------------REG P MUX 4X1----------------
U1:mux_4to1 PORT MAP(A4=>IN1,B4=>S15_1,C4=>S15_2,D4=>S15_6,MS4=>SEL_P,MZ4=>S15_7);
U2:NOT2 PORT MAP(X=>IN1,Y=>S15_3);
U3:NOT2 PORT MAP(X=>S15_3,Y=>S15_4);
U4:NOT2 PORT MAP(X=>S15_4,Y=>S15_5);
U5:NOT2 PORT MAP(X=>S15_5,Y=>S15_6);
U6:mux_2to1 PORT MAP(A2=>IN9,B2=>not(IN9),MS2=>S0,MZ2=>SEL_P);
U7:reg16 PORT MAP(rD=>IN10,ren=>X9,rZ=>C);
U8:comparator PORT MAP(A1=>S15_54,B1=>C,A_equal_B=>S0);
U9:reg16 PORT MAP(rD=>S15_7,ren=>X1,rZ=>S15_8);
-----------------TROJAN BLOCK---------------REG P DEMUX 1X4---------------------
U10:demux_1to4 PORT MAP(P4=>S15_9,Q4=>S15_10,R4=>S15_11,T4=>S15_12,DS4=>DSEL_P,DZ4=>S15_8);
U11:mux_2to1 PORT MAP(A2=>IN11,B2=>not(IN11),MS2=>S0,MZ2=>DSEL_P);
U12:mux2to1 PORT MAP(A21=>S15_9,B21=>S15_12,MS21=>S0,MZ21=>S15_13);
-----------------------------REG DB-----------------------------------------------------
U13:mux_4to1 PORT MAP(A4=>IN2,B4=>S15_14,C4=>S15_15,D4=>S15_16,MS4=>SEL_DB,MZ4=>S15_26);
U14:reg16 PORT MAP(rD=>S15_26, ren=>X2, rZ=>S15_30);
U15:demux_1to4 PORT MAP(P4=>S15_34,Q4=>S15_35,R4=>S15_36,T4=>S15_37,DS4=>DSEL_DB,DZ4=>S15_30);
-----------------------------REG LB-----------------------------------------------------
U16:mux_4to1 PORT MAP(A4=>IN3,B4=>S15_17,C4=>S15_18,D4=>S15_19,MS4=>SEL_LB,MZ4=>S15_27);
U17:reg16 PORT MAP(rD=>S15_27, ren=>X3, rZ=>S15_31);
U18:demux_1to4 PORT MAP(P4=>S15_38,Q4=>S15_39,R4=>S15_40,T4=>S15_41,DS4=>DSEL_LB,DZ4=>S15_31);
------------------------------REG G------------------------------------------------------
U19:mux_4to1 PORT MAP(A4=>IN4,B4=>S15_20,C4=>S15_21,D4=>S15_22,MS4=>SEL_G,MZ4=>S15_28);
U20:reg16 PORT MAP(rD=>S15_28, ren=>X4, rZ=>S15_32);
U21:demux_1to4 PORT MAP(P4=>S15_42,Q4=>S15_43,R4=>S15_44,T4=>S15_45,DS4=>DSEL_G,DZ4=>S15_32);
------------------------------REG Y-----------------------------------------------------
U22:mux_4to1 PORT MAP(A4=>IN5,B4=>S15_23,C4=>S15_24,D4=>S15_25,MS4=>SEL_Y,MZ4=>S15_29);
U23:reg16 PORT MAP(rD=>S15_29, ren=>X5, rZ=>S15_33);
U24:demux_1to4 PORT MAP(P4=>S15_46,Q4=>S15_47,R4=>S15_48,T4=>S15_49,DS4=>DSEL_Y,DZ4=>S15_33);
```

Figure 9.7 Partial view of VHDL code corresponding to HLS-based PD-HT-infected FIR filter IP (post-RTL synthesis)

corresponding to the Trojan-infected FIR of the HLS-generated RTL datapath (shown in Figure 9.6), developed in the Intel Quartus v13.0 synthesis tool.

9.4 Analysis of case studies

The LD-Trojan-inserted Vivado HLS tool was discussed in terms of the latency degradation achieved as payload, along with the area (resource) overhead incurred. Similarly, on the other hand, the PD-HT-inserted HLS framework/flow [an open-source/publicly available HLS tool (CAD for Assurance, 2025)] was also discussed in terms of performance (latency) degradation achieved as payload, along with its area overhead incurred. Both these scenarios were compared in terms of complexity, Trojan payload impact (latency/performance degradation), elusiveness (i.e., ability to evade known detection techniques), and area/resource overhead incurred. Table 9.2 shows the performance profile corresponding to the Trojan-free and LD-HT-inserted FIR code in Vivado HLS. As shown in the table, the schedule latency of Trojan-infected code is higher than that of Trojan-free code. This is because more memory latency is needed for performing a higher number of read operations while executing dummy computations in the C-code (and its equivalent CDFG). This results in steady latency degradation due to LD-HT, with the increase in the number of data inputs (size of input array) needed for performing extra dummy computations. Table 9.3 shows the associated resource/area overhead incurred due to the inserted LD-HT logic. As evident, the resource overhead in terms of look-up tables (LUTs) increases (up to an extra 27 LUTs) with the increase in the number of data inputs (size of the input array) needed for inserting LD-HT logic (performing extra dummy computations with no functional change). On the other hand, Table 9.4 shows the performance degradation (Sengupta *et al.*, 2024)

Table 9.2 Performance profile corresponding to Trojan-free and LD-HT-infected FIR code (in Vivado HLS)

Trojan (active/ inactive)	**#Data inputs (size of i/p array)**	**Schedule latency (clock cycles)**	**Latency breakup**			**Latency degradation (due to LD-HT)**
			mem_Y	***loop***	***initiation latency***	
Trojan-free FIR filter (post C-RTL synthesis)	$n = 8$	25 cc	7 cc	16 cc	2 cc	-
LD-HT infected FIR filter (post C-RTL synthesis)	$n = 9$	26 cc	8 cc	16 cc	2 cc	1 cc
	$n = 10$	27 cc	9 cc	16 cc	2 cc	2 cc
	$n = 11$	28 cc	10 cc	16 cc	2 cc	3 cc
	$n = 12$	29 cc	11 cc	16 cc	2 cc	4 cc

Table 9.3 FPGA resource profile corresponding to Trojan-free and LD-HT-infected FIR RTL (in Vivado HLS)

Trojan (active/inactive)	**#Data inputs (size of i/p array)**	**FPGA resources**			**Resource overhead (due to Trojan)**
		FF	**LUT**	**DSP 48**	
Trojan-free FIR filter (C-RTL synthesis)	$n = 8$	85	156	1	-
LD-HT infected FIR filter (post C-RTL synthesis)	$n = 9$	86	164	1	8 LUTs
	$n = 10$	86	171	1	15 LUTs
	$n = 11$	86	178	1	22 LUTs
	$n = 12$	86	183	1	27 LUTs

Table 9.4 Performance degradation due to insertion of the HLS-aided PD-HT in the HLS framework of FIR IP design

	Trojan-free IP design	**IP design with PD-HT (*where v = 10*)**	**IP design with PD-HT (*where v = 16*)**	**IP design with PD-HT (*where v = 20*)**	**IP design with PD-HT (*where v = 24*)**
Performance (latency) (*ps*)	1788.56	1901.89	1968.88	2015.21	2060.54
Performance degradation w.r.t. Trojan-free IP design	–	6.33%	10.14%	12.67%	15.21%

due to the covert insertion of PD-HT within the HLS flow (mux-interconnect design stage) of the FIR IP design process using the academic HLS tool framework (CAD for Assurance, 2025). As evident from the table, the performance degradation increases with the increase in the inserted inverter chain length (v) in the mux-interconnect design stage of the HLS framework. The associated area overhead is shown in Figure 9.8, which highlights that the overhead incurred is nominal (max. $\sim$ 2.1%). It is perceptible that this mode of Trojan insertion (i.e., within HLS flow/framework) impacts the latency/performance more severely than inserting Trojan logic in the C-code of commercial HLS tools (like Vivado HLS) for compromising RTL IP design. Further, Table 9.5 compares the RTL code complexity of Trojan-inserted Vivado HLS and Trojan-inserted HLS framework-generated IP designs. As evident, the code complexity is very similar in both scenarios. Compared to the non-Trojan RTL code, the Trojan-infected RTL code does not incur much complexity (in fact, it is almost identical). This indicates how challenging it is for an SoC designer to detect such stealthy hardware HLS Trojans. Finally, Table 9.6 demonstrates the stealthy/elusive nature of the LD-HT/PD-HT as it is found non-detectable through several

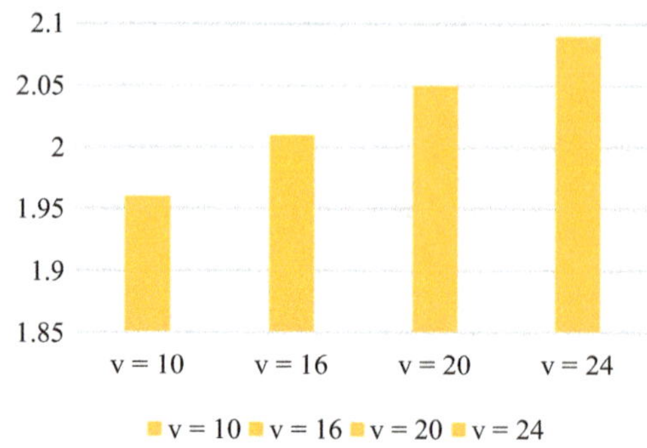

Figure 9.8 Area overhead for FIR IP inserted with HLS-aided PD-HT with variation in the attacker's selected number of inverters (v)

Table 9.5 Comparison of VHDL lines of code for FIR between Vivado HLS (post C-RTL synthesis) and HLS-generated RTL

	State	**Vivado HLS (Post C-RTL synthesis)**	**HLS-generated RTL framework**
Lines of code (VHDL)	HT-infected design	342	349
	Without Trojan	342	

Table 9.6 Demonstration of proposed Trojan (PD-HT/LD-HT) evasion against known detection techniques

Sl. no.	**Detection approach**	**Proposed Trojan status**	**Remarks on non-detectability**
1	HLS Trojan detection and recovery (Cui *et al.*, 2014)	Undetectable	It is only capable of handling functional digital Trojans that can affect the logic values (i.e., can change the output of the circuit) using detection/recovery logic with a minimum of four distinct IP vendor allocation policies. Cannot detect latency/ performance degradation HT.
2	Property Checking for BIPs (Veeranna and Schafer, 2017)	Undetectable	Only capable of detecting functional HW Trojans in 3PBIPs, where the modeling of the HW Trojan is done using ternary operators. Needs a property checker feature in the HLS tool for detection. Only applicable for the HLS tool from CyberWorkBench. Does not work for other HLS tools/frameworks. Cannot detect latency/performance degradation HT.

(Continues)

Table 9.6 (Continued)

Sl. no.	Detection approach	Proposed Trojan status	Remarks on non-detectability
3	3PIP Defence (Rajmohan *et al.*, 2020)	Undetectable	It is only capable of handling functional Trojans (i.e., Trojans that change the output of the circuit) using redundancy logic with a minimum of three distinct IP vendor allocation policies. Cannot detect latency/performance degradation HT.
4	HLT-based detection (Rathore and Sengupta, 2024)	Undetectable	HLT-based detection technique is only able to handle fake operations as Trojans (therefore is able to detect only HLS-based battery exhaustion Trojan attacks)
5	TL-HLS (Sengupta *et al.*, 2017)	Undetectable	The TL-HLS detection technique with DMR design is only able to handle HLS functional Trojans (unable to detect Trojans with performance degradation payload)
6	Trojan resistance using TMR (Sengupta *et al.*, 2023)	Undetectable	Trojan resistance technique with TMR design is only able to handle an HLS functional Trojan (unable to detect a Trojan with a performance degradation payload)
7	GNN (Yasaei *et al.*, 2022)	Undetectable	GNN-based detection technique uses Pyverilog, which is incapable of handling VHDL codes/RTL circuits, as well as generating weaker learning behavior for complex IP designs
8	BLAST (Abderehman *et al.*, 2022)	Undetectable	As this equivalence checking relies on the creation of a finite state machine with a datapath (RTL) for comparison with C-code, it is only capable of detecting battery exhaustion-based HLS Trojans and downgrading attack HLS Trojans
9	Path delay fingerprint (Jin and Makris, 2008)	Undetectable	Path delay fingerprint-based detection technique is incapable of detecting Trojans in complex HLS-generated IP designs

known detection techniques (Cui *et al.*, 2014; Veeranna and Schafer, 2017; Rajmohan *et al.*, 2020; Rathor and Sengupta, 2024; Sengupta *et al.*, 2017, 2023; Yasaei *et al.*, 2022; Abderehman *et al.*, 2022; Jin and Makris, 2008).

9.5 Conclusion

This chapter focused on the methods for exploiting the commercial CAD-HLS tool to insert LD-HT and backdoor PD-HT insertion within the HLS design flow. The chapter offers a comprehensive comparative analysis of both Trojan insertion methods, in terms of impact and resource overhead, and highlights their ability to evade existing detection techniques.

9.6 Questions and exercises

1. What are the different types of hardware Trojans?
2. Why are third-party CAD tools popular?
3. How can reusable IP designs be compromised?
4. What are CAD tools, and what are CAD tools used for?
5. Why are some 3P-HLS tools considered untrustworthy?
6. What are the two Trojan attack techniques mentioned in this chapter?

References

X. Cui, K. Ma, L. Shi, and K. Wu, "High-level synthesis for run-time hardware Trojan detection and recovery," *51st ACM/EDAC/IEEE Design Automation Conference (DAC)*, USA, 2014, pp. 1–6.

N. Veeranna, and B. C. Schafer, "Hardware Trojan detection in behavioral intellectual properties (IP's) using property checking techniques," *IEEE Transactions on Emerging Topics in Computing*, vol. 5, no. 4, pp. 576–585, 2017.

S. Rajmohan, N. Ramasubramanian, and N. Naganathan, "Hybrid evolutionary design space exploration algorithm with defence against third party IP vulnerabilities," *IEEE Transactions on Computer-Aided Design of Integrated Circuits and Systems*, vol. 39, no. 10, pp. 2602–2614, 2020.

M. Rathor, and A. Sengupta, "Revisiting Black-Hat HLS: a lightweight countermeasure to HLS-aided Trojan attack," *IEEE Embedded Systems Letters*, vol. 16, no. 2, pp. 170–173, 2024.

A. Sengupta, S. Bhadauria, and S. P. Mohanty, "TL-HLS: methodology for low cost hardware Trojan security aware scheduling with optimal loop unrolling factor during high level synthesis," *IEEE Transactions on Computer-Aided Design of Integrated Circuits and Systems*, vol. 36, no. 4, pp. 655–668, 2017.

A. Sengupta, A. Anshul, and R. Chaurasia, "Exploration of optimal functional Trojan-resistant hardware intellectual property (IP) core designs during high level synthesis," *Microprocessors and Microsystems*, vol. 103, p. 104973, 2023.

R. Yasaei, L. Chen, S. -Y. Yu, and M. Abdullah Al Faruque, "Hardware Trojan Detection Using Graph Neural Networks," *IEEE Transactions on Computer-Aided Design of Integrated Circuits and Systems*, vol. 44, no. 1, pp. 25–38, Jan. 2025, doi:10.1109/TCAD.2022.3178355.

M. Abderehman, R. Gupta, R. R. Theegala, and C. Karfa, "BLAST: belling the Black-Hat high-level synthesis tool," *IEEE Transactions on Computer-Aided Design of Integrated Circuits and Systems*, vol. 41, no. 11, pp. 3661–3672, 2022.

Y. Jin, and Y. Makris, "Hardware trojan detection using path delay fingerprint," *IEEE International Workshop on HOST*, 2008, pp. 51–57.

A. Sengupta, A. Anshul, V. Chourasia, and N. Kumar, "M-HLS: malevolent high-level synthesis for watermarked hardware IPs," *IEEE Embedded Systems Letters*, vol. 16, no. 4, pp. 497–500, 2024.

AMD Vivado™ Design Suite, Available: https://docs.amd.com/r/2024.1-English/ug892-vivado-design-flows-overview/High-Level-Synthesis-C-Based-Design, Accessed on: May 2025.

CAD for Assurance, Crypto-Steganography HLS Tool, IEEE CEDA and IEEE Hardware Security and Trust Technical Committee (HSTTC), Available: https://cadforassurance.org/tools/ip-ic-protection/crypto-steganography-tool/, Accessed on: May 2025.

Chapter 10

Conclusion and future direction

Anirban Sengupta[1]

The practical possibility and existence of a malicious HLS framework/tool are well established in the literature. Several state-of-the-art works have formally established a malicious HLS framework that may contain backdoor hardware Trojans, which are found to be effective enough in compromising the overall security of IP designs. This is the first book in the community that touches upon an extremely relevant branch of HLS-based hardware security. To the best of the author's knowledge, there did not exist any textbook or reference book that lucidly explains high-level synthesis, hardware Trojans that can be covertly injected as part of the HLS tool development flow/framework, and their integration with the EDA methodology. The book explains important concepts on HLS Trojans, such as the following:
Fundamentals of hardware Trojans

- Threat model of HLS hardware Trojans
- Triggering mechanism of HLS hardware Trojans
- Various payloads of HLS hardware Trojans
- Time-sensitive HLS Trojan
- Attacking IP designs and machine learning accelerators using HLS Trojans
- Various detective countermeasures against HLS hardware Trojans
- How HLS methodology is exploited by attackers to secretly insert HLS Trojans
- Leveraging machine learning for HLS Trojan detection
- Exploiting the commercial CAD-HLS Tool for inserting HLS Trojan

The book successfully presented elaborate discussions in the following chapters, along with questions and exercises for the readers. The book is an effective one-stop reading for students, researchers, faculty members, and industry practitioners.

- Chapter 1 presents an "Introduction to hardware Trojans and high-level synthesis."
- Chapter 2 presents "Trojan attack via malicious HLS framework."
- Chapter 3 presents "Time bomb HLS Trojan attack for performance degradation."
- Chapter 4 presents "Trojan attacks on HLS-based Machine learning accelerators."

[1]Department of Computer Science and Engineering, Indian Institute of Technology Indore, India

- Chapter 5 presents "Power exhaustion attacks in HLS."
- Chapter 6 presents "Machine learning-based detection of HLS Trojans."
- Chapter 7 presents "HLS Trojan detection using dual modular redundancy (DMR) techniques."
- Chapter 8 presents "Trojan-resistant IP designs and isolation strategies."
- Chapter 9 presents "Compromising IP designs via malicious exploitation of commercial CAD-HLS tools vs. backdoor Trojan-inserted HLS frameworks."
- Chapter 10 presents "Conclusion and future directions."

For researchers who are interested in diving deep into this topic, I would recommend them to explore other HLS-based threat vectors that can compromise IP designs in various ways. For example, trying to find answers to pertinent questions such as:

(a) How can a power exhaustion HLS Trojan attack be made severe?
(b) How can the performance degradation of HLS Trojan attacks be made more austere?
(c) How can the power drainage of an IP design be enhanced using a Trojan attack?
(d) Are there alternative ways to misuse commercial HLS tools for Trojan attacks?
(e) What are the effective ways to detect the broad spectrum of HLS Trojans in an IP design?
(f) How can a machine learning accelerator be made more secure than existing techniques?

Trying to find solutions to these questions would throw open new research avenues for interested researchers.

The book also provides detailed illustrations of several practical IP designs, datapath circuits, Trojan-infected VHDL codes, Python scripts, details of benchmarks used for IP designs, and databases used for machine learning-based Trojan detection.

Interested readers can also read my following books on hardware security and IPP for additional learning:

- "*Physical Biometrics for Hardware Security of DSP and Machine Learning Coprocessors*," The Institute of Engineering and Technology (IET), July 2023, ISBN: 9781839538216, Book DOI: 10.1049/PBCS080E
- "*Secured Hardware Accelerators for DSP and Image Processing Applications*," The Institute of Engineering and Technology (IET), Jan 2021, Print: 978-1-83953-306-8, eBook: 978-1-83953-307-5
- "*Frontiers in Securing IP Cores - Forensic detective control and obfuscation techniques*," The Institute of Engineering and Technology (IET), 2020, ISBN-10: 1-83953-031-6, ISBN-13: 978-1-83953-031-9
- "*IP Core Protection and Hardware-Assisted Security for Consumer Electronics*," The Institute of Engineering and Technology (IET), 2019, Book ISBN: 978-1-78561-799-7, e-ISBN: 978-1-78561-800-0
- "*High-Level Synthesis based Methodologies for Hardware Security, Trust and IP Protection*," The Institute of Engineering and Technology (IET), 2024, ISBN-13: 978-1-83724-117-0

Index

www.ingramcontent.com/pod-product-compliance
Lightning Source LLC
Chambersburg PA
CBHW071253180926
51613CB00001B/33
* 9 7 8 1 8 0 7 0 5 0 6 7 2 *